Applying Pharmacogenomics to Optimize Geriatric Care

Teresa DeLellis • Yee Ming Lee
David Bright • Youssef Roman
Stephanie Sibicky
Editors

Applying Pharmacogenomics to Optimize Geriatric Care

Springer

Editors
Teresa DeLellis
Chair, Department of Pharmacy Practice
Manchester University, College of Health Professions, Nursing & Pharmacy
Fort Wayne, IN, USA

David Bright
Professor of Pharmaceutical Sciences
College of Pharmacy, Ferris State University
Big Rapids, MI, USA

Stephanie Sibicky
Clinical Professor and Assistant Dean for Academic Affairs and Assessment
Bouvé College of Health Sciences, Northeastern University
Boston, MA, USA

Yee Ming Lee
Associate Professor of Clinical Pharmacy
Skaggs School of Pharmacy and Pharmaceutical Sciences, University of Colorado
Aurora, CO, USA

Youssef Roman
Associate Professor of Pharmacy Practice and Administrative Sciences
S. Skaggs College of Pharmacy
Boise, ID, USA

Editorial Contact: Jessica Chio

ISBN 978-3-032-10652-0 ISBN 978-3-032-10653-7 (eBook)
https://doi.org/10.1007/978-3-032-10653-7

This Springer imprint is published by the registered company Springer Nature Switzerland AG
The registered company address is: Gewerbestrasse 11, 6330 Cham, Switzerland

Preface

No matter the field or discipline, almost every healthcare professional has undoubtedly heard their patients say something like "well, that just runs in my family," "my mother/father also had that," or "I come by it honestly." Many clinicians have long considered family history when weighing disease risk and treatment decisions, with clinical guidelines taking family history into account to varying degrees for many health conditions. Until recently, it was far less clear how family history of medication response could be properly considered in the context of medication treatment decisions. However, the last two decades have spurred significant advances in application of pharmacogenetics (PGx), or the study of how genetic variations influence medication response. Not only has laboratory testing expanded greatly to objectively identify genetic variants responsible for medication response, but clinical guidelines have also been released and revised to assist healthcare professionals in drug selection and dosing. As the science of precision medicine, genetic testing, and clinical guidance expand rapidly, application of PGx is increasingly becoming an expectation in many areas of health care.

While PGx adds key information for individualizing therapy to achieve targeted treatment responses and minimize adverse effects, few populations stand to benefit more than older adults. Older adults consume the most medications while being at the highest risk for both medication adverse events and treatment failures. In addition, the older adult population is also highly complex, offering the greatest possibility of phenoconversion (defined as a mismatch between genotype-predicted phenotype and the observed phenotype). The most common and well-understood cause of phenoconversion is drug-drug interactions. This phenomenon has also been described as the combination of drug-drug-gene interactions (when drug-drug interactions and drug-gene interactions coincide). Phenoconversion obfuscates the best course of action to take based on PGx data, and few trials conducted in older adults exist to guide the application of PGx in this complex patient population.

This text is designed for any healthcare provider who cares for older adults and wishes to advance their understanding of PGx application in geriatrics. Through a case-based approach and examination of existing literature, this text provides additional tools to navigate the complexity of caring for older adults to provide evidence-based and personalized treatment decisions. Part I reviews the most salient principles in both geriatrics and PGx necessary to digest the application portion of the text, ending with a summary of what is known about the intersection of these two clinical and scientific disciplines. Part II contains chapters of case applications to inform common scenarios encountered in the areas that overlap between geriatrics and PGx. The goal of this text, through the approach outlined below, is to help readers create a framework for approaching these situations individually, considering the lack of direct evidence in the intersection of these realms, and synthesizing concepts from Section I to guide treatment decisions.

The proposed framework, as followed in the case discussions in Section II, essentially boils down to this:

1. Identify the clinical, nongenetic factors influencing geriatric care and the associated guidance relevant to question at hand (effects of aging physiology, multimorbidity, and polypharmacy on treatment options). This includes consulting evidence-based and holistic care guidelines and other resources, such as the AGS Beers Criteria® and published guidance on pharmacologic and nonpharmacologic alternatives, to compile a list of reasonable treatment options. Care beyond medications, including interprofessional team consultations, should also be applied.
2. If PGx information is available, consider guideline-based dosing and selection guidance for each of the treatment options arrived at through Step 1.
3. Finally, review the information from Step 1 considerations for the possibility of phenoconversion. Consider modifying the treatment recommendations from the completion of Steps 1 and 2 based on any reasonable possibility of phenoconversion.

It is important to recognize that PGx data does not always lead to changes in medication therapy decisions. In some instances, PGx information may be completely irrelevant—particularly when a medication shows no known association between genetic variation and drug response. In other cases, PGx data may support continued use of a medication by reinforcing its expected efficacy, thereby increasing clinical confidence rather than prompting a change. PGx may also assist with overcoming clinical inertia in deprescribing medications for older adults by underscoring the likelihood of eventual adverse effects or low likelihood of benefit, especially in cases where true efficacy is difficult to ascertain. There are also scenarios where patients respond well to medications with PGx implications, underscoring that PGx is just one component of a broader picture. There may be other genetic factors not captured in the current PGx analysis that may influence medication response. This text will highlight examples where PGx data should and should not inform medication selection or dosing in older adults, with the goal of enhancing medication efficacy and minimizing adverse effects.

Each chapter in Section II begins with a brief introduction highlighting key geriatric and PGx concepts relevant to the chapter's organ system. This is followed by case studies that demonstrate how to apply a standardized approach to interpreting and integrating PGx information within the context of clinical geriatric care. The cases are intentionally streamlined, with limited patient detail, as the purpose of this text is not to provide an exhaustive review of geriatric pharmacotherapy. Instead, the goal is to support clinicians—particularly those with some experience with geriatrics—in building confidence and practical skills in applying PGx to real-world scenarios. In clinical practice, there is typically extensive information in medical records for older adults. Rather than replicating this complexity, the cases focus on the most critical elements that influence therapeutic decisions. The accompanying discussions acknowledge that additional clinical data may be available or accessible, while maintaining a targeted emphasis on optimizing therapy through PGx insights.

Each case concludes with a summary that includes clinical recommendations. These recommendations are intentionally open-ended, reflecting the limited patient-specific data provided and acknowledging the variability in real-world scenarios, clinician preferences, local prescribing practices, and other nuances inherent to the diverse older adult populations. The recommendations also highlight how care management may differ when PGx information is considered, emphasizing its potential impact on therapeutic decisions.

Fort Wayne, IN, USA — Teresa DeLellis
Aurora, CO, USA — Yee Ming Lee
Big Rapids, MI, USA — David Bright
Boise, ID, USA — Youssef Roman
Boston, MA, USA — Stephanie Sibicky

Contents

Part I
Foundations

Chapter 1
Physiologic Effects of Aging, Multimorbidity, and Polypharmacy on Care Outcomes

Teresa DeLellis and Stephanie Sibicky

Caring for older adults requires a comprehensive understanding of the physical, environmental, social, and economic challenges that impact the use of medications in this population. Compared to younger adults, older adults experience multimorbidity [3], medication adverse events, and hospitalizations at higher rates [6]. Optimizing medication therapy for older adults is inherently complex, requiring the balance of several conditions with their associated lists of medications, the risk and benefit of adverse effects and prognosis, and understanding the patient's goals of care and quality of life concerns. Three critical aspects must be weighed when designing a medication regimen for an older adult: physiological changes associated with aging, multimorbidity, and polypharmacy.

Human physiology is not static. It undergoes constant dynamic change from birth to the end of life. A few key examples of physiological changes that occur with aging are listed in Table 1.1. As each organ system and tissue evolves with age, the way medications are processed by the body and the effect they have on the body also change. Examples of pharmacokinetic changes—changes in how the body processes medications—are listed in Table 1.2. These changes necessitate adaptation in how medications are prescribed and dosed to (1) reduce the risk of compounded medication accumulation, which could lead to an increased risk for serious adverse effects, and (2) adjust to the changes in absorption, distribution, and metabolism that could lead to treatment failures.

T. DeLellis (✉)
School of Pharmacy, College of Health Professions, Nursing and Pharmacy, Manchester University, Fort Wayne, IN, USA
e-mail: TMDeLellis@manchester.edu

S. Sibicky
Clinical Professor and Assistant Dean for Academic Affairs and Assessment, Bouvé College of Health Sciences, Northeastern University, Boston, MA, USA

T. DeLellis et al. (eds.), *Applying Pharmacogenomics to Optimize Geriatric Care*, https://doi.org/10.1007/978-3-032-10653-7_1

Table 1.1 Age-related physiologic changes

System	Key changes
Neurologic	↓ brain mass, ↓ cerebral blood flow & autoregulation, ↓ neurotransmitters (dopamine, serotonin), ↓ receptors, ↓ sleep efficiency, ↑ comorbidities (depression, dementia, Parkinson's disease)
Cardiovascular	↑ arterial stiffness, ↑ orthostasis (↓ baroreceptor sensitivity), ↑ sympathetic activity (↓ beta/alpha responsiveness), endothelial dysfunction, low renin activity
Respiratory	↓ thoracic cavity size (muscle & vital capacity changes), ↓ reserve, ↓ cough strength (↓ clearance), impaired immunity (↑ infection risk), ↓ lung elasticity, ↑ residual volume, ↓ FEV1
Gastrointestinal tract	↓ saliva, dental trauma (dentures, more dental work), xerostomia (meds/comorbidities), taste disturbances, slowed peristalsis, ↓ esophageal sphincter pressures, ↑ aspiration risk, dysphagia/odynophagia, ↑ stomach pH, loss of mucosal glands (→ hypo−/achlorhydria), ↑ ulcer risk, atrophic gastritis (↓ pain receptors), ↓ intestinal mucosal surface, small intestine mostly unchanged (meds/comorbidities may alter), altered microbiome → inflammation & ↓ protective mechanisms
Kidney/renal	↓ creatinine production (muscle loss), ↓ renal blood flow, ↓ nephrons/glomeruli/tubules/renal mass, impaired tubular secretion, impaired urine concentration (↑ dehydration risk), ↓ GFR

Table 1.2 Age-related pharmacokinetic changes

Category	Key changes
Absorption	Altered GI acidity (↓ acid → ↓ absorption e.g., iron), slowed gastric motility (delayed absorption, altered site absorption), ↓ first-pass metabolism → ↑ bioavailability (↑ plasma concentrations); ↓ prodrug activation
Distribution	↓ total body water, ↓ lean muscle mass, ↑ total body fat → ↑ volume of distribution for lipophilic drugs (accumulation risk); water-soluble drugs ↑ concentrations (toxicity risk); ↓ albumin in frailty → ↑ free drug (especially protein-bound drugs)
Metabolism	↓ liver mass and hepatic blood flow → ↓ metabolism of low extraction drugs (slower clearance), ↑ metabolism of high extraction drugs (dose adjustments needed); variable CYP450 & transporter (p-glycoprotein) changes
Elimination	↓ kidney mass and renal blood flow → ↓ clearance of renally excreted drugs (dose adjustments needed), impaired biliary excretion (gallbladder/liver changes), altered clearance pathways → risk of toxicity or therapeutic failure

Pharmacodynamic changes, the effect medications have on the body, are less well defined but are also affected by aging physiology. The most common and impactful changes to pharmacodynamics involve the cardiovascular and central nervous systems. Orthostasis, QT interval prolongation, and isolated systolic hypertension become more common and present treatment challenges, while some key medications, such as beta-blockers, become less effective due to receptor downregulation. The blood–brain barrier becomes more permeable with aging, and the reserve of neurotransmitters is decreased in the central nervous system (CNS). This translates into an increased risk of CNS-related adverse effects, especially cognitive, vestibular, and motor effects, and an increased risk of treatment failure for medications with a neurotransmitter-based mechanism of action.

As a person ages, the chance of developing illnesses and accumulating medications also increases. For many older adults, particularly those prioritized for pharmacogenomic (PGx) testing, this may lead to multimorbidity (defined as the presence of two or more chronic conditions) and polypharmacy (defined as taking five or more prescription medications) [3, 4]. Each disease state can interact with others to influence physiological effects, health outcomes, and medication reactions (e.g., treatment outcomes, adverse effects). For example, treating an older adult with orthostatic hypotension and Parkinson's disease must balance the risk of falls due to low blood pressure and the movement difficulties as a nature of their Parkinson's. Medication interactions (commonly referred to as drug–drug interactions) are also increasingly common and complex in the setting of polypharmacy, as drug–drug interactions layer upon each other making it increasingly difficult to predict the ultimate result in terms of efficacy and safety of medications. For example, if an older adult requires the administration of amiodarone as a rate or rhythm control agent for their atrial fibrillation, it can interact with the anticoagulation (e.g., warfarin, apixaban) necessary to prevent stroke from atrial fibrillation. The final layer is disease states and medications interacting with each other, often referred to as drug-disease interactions. An example of this would be an older adult taking an anticholinergic medication for their urinary incontinence, then experiencing worsened constipation as an adverse effect. Treating the patient's constipation as a result of the medication's adverse effect is often described as the "prescribing cascade," where one medication is added to the patient's regimen to treat the adverse effect of another. When one combines the effects of multimorbidity and polypharmacy interactions, layered with the additional complexities of drug-disease interactions, the result is the ultimate level of what is now referred to as "multicomplexity" in the field of geriatrics [8].

All these factors (physiologic changes of aging, pharmacokinetic and pharmacodynamic changes, multimorbidity, and polypharmacy) interact with each other to make each older adult patient infinitely unique. These overlapping influences contribute to the development of geriatric syndromes, such as frailty, delirium, and falls, which reflect the complex interplay of multiple health and functional domains rather than a single disease. The culmination of these can also lead to the development of new adverse effects emerging after years of stability on a medication. It is no surprise that treatment failure and adverse effect rates are high, considering how challenging it is to accurately predict the effect of medications, given the lack of evidence to guide clinical decision-making in the setting of multicomplexity. Making matters worse, even after controlling for such key differences in patients, response to treatment can still vary drastically given other unique patient characteristics, such as their genetic makeup. PGx, explored in depth in the following chapter, is one resource with the primary aim of decreasing treatment failures and adverse effects by identifying a person's genetic variation to eliminate one "unknown" factor that influences these outcomes [7].

Considering these treatment challenges, the approach to medication use in older adults should differ dramatically from that used to treat younger adults. Multimodal strategies, emphasizing non-pharmacologic treatments and leveraging interprofessional teams, are at the core of minimizing high-risk medication use, addressing social determinants of health, and optimizing care outcomes. When medications are necessary, there are several clinical decision support tools available that should be routinely implemented, especially the Medication Appropriateness Index, the AGS Beers Criteria®, and guidance for alternative treatments to high-risk medications [1, 2, 5].

References

1. 2003 American Geriatrics Society Beers Criteria® Update Expert Panel. American Geriatrics Society 2023 updated AGS Beers Criteria® for potentially inappropriate medication use in older adults. J Am Geriatr Soc. 2023;71(7):2052–81. https://doi.org/10.1111/jgs.18372. Epub 2023 May 4. PMID: 37139824
2. American Geriatrics Society Beers Criteria® Alternatives Panel, Steinman MA, Semla TP, Akgün KM, et al. Alternative treatments to selected medications in the 2023 American Geriatrics Society Beers Criteria®. J Am Geriatr Soc. 2025:1–21. https://doi.org/10.1111/jgs.19500.
3. Chowdhury SR, Chandra Das D, Sunna TC, Beyene J, Hossain A. Global and regional prevalence of multimorbidity in the adult population in community settings: a systematic review and meta-analysis. EClinicalMedicine. 2023;57:101860. https://doi.org/10.1016/j.eclinm.2023.101860. PMID: 36864977; PMCID: PMC9971315
4. Hales CM, Servais J, Martin CB, Kohen D. Prescription drug use among adults aged 40–79 in the United States and Canada. NCHS data brief, no 347. Hyattsville: National Center for Health Statistics; 2019.
5. Hanlon JT, Schmader KE, Samsa GP, Weinberger M, Uttech KM, Lewis IK, Cohen HJ, Feussner JR. A method for assessing drug therapy appropriateness. J Clin Epidemiol. 1992;45(10):1045–51. https://doi.org/10.1016/0895-4356(92)90144-c. PMID: 1474400
6. Oscanoa TJ, Lizaraso F, Carvajal A. Hospital admissions due to adverse drug reactions in the elderly. A meta-analysis. Eur J Clin Pharmacol. 2017;73(6):759–70. https://doi.org/10.1007/s00228-017-2225-3. Epub 2017 Mar 1. PMID: 28251277
7. Roman YM, Dixon DL, Salgado TM, et al. Challenges in pharmacotherapy for older adults: a framework for pharmacogenomics implementation. Pharmacogenomics. 2020;21(9):627–35. https://doi.org/10.2217/pgs-2019-0198.
8. Tinetti M, Huang A, Molnar F. The geriatrics 5M's: a new way of communicating what we do. J Am Geriatr Soc. 2017;65(9):2115. https://doi.org/10.1111/jgs.14979. Epub 2017 Jun 6. PMID: 28586122

Chapter 2
Introduction to Pharmacogenomics

David Bright, Youssef Roman, and Yee Ming Lee

Introduction

Most health care professionals can attest to observing a wide variation in medication response among patients. As highlighted in Chap. 1, older adults often respond to medications differently than younger patients. Adjusting for factors such as age, weight, sex, renal function, and other patient-specific characteristics is a well-established practice in medicine. Despite these adjustments, variability in medication response may still occur. Pharmacogenomics (PGx) offers new opportunities to further personalize medication therapy by integrating an individual's genetic information into clinical decision-making to help transform the traditional approach to medicine.

Pharmacogenomics examines how an individual's genetic makeup influences their response to medications. Many clinicians find it challenging to digest PGx reports and integrate them into comprehensive care plans given the time constraints present in clinical practice. In addition to implementing the resources and approach outlined in this text, PGx is another area where leveraging interprofessional teams, as highlighted in Chap. 1, is a key strategy for providing holistic care for older

D. Bright (✉)
Professor of Pharmaceutical Sciences, College of Pharmacy, Ferris State University, Big Rapids, MI, USA
e-mail: davidbright@ferris.edu

Y. Roman
L.S. Skaggs College of Pharmacy, Idaho State University, Pocatello, ID, USA
e-mail: youssefroman@isu.edu

Y. M. Lee
Associate Professor of Clinical Pharmacy, Skaggs School of Pharmacy and Pharmaceutical Sciences, University of Colorado, Aurora, CO, USA
e-mail: yeeming.lee@cuanschutz.edu

T. DeLellis et al. (eds.), *Applying Pharmacogenomics to Optimize Geriatric Care*, https://doi.org/10.1007/978-3-032-10653-7_2

adults. Given their extensive training in pharmacology and therapeutics, pharmacists are well-positioned to address PGx-related considerations across various care settings [9]. Through collaboration with the broader healthcare team, pharmacists play a critical role in integrating patient-specific PGx data to optimize medication therapy and enhance patient outcomes.

Key Genetic Terminology

While a comprehensive review of human genetics is beyond the scope of this textbook, the following are some commonly encountered genetic terms in the PGx literature (see Table 2.1) [14]. To support consistent PGx data sharing across diverse electronic health record systems, many laboratories have adopted standardized PGx terms in their reports (see Table 2.2) [2].

Table 2.1 List of genetic terms and their definition

Genetic term	Definition
Allele	An alternative form of a gene or DNA sequence
Diplotype	The combination of star alleles inherited from each parent for a pharmacogene, e.g., CYP2C19*1/*2 diplotype shows *1 allele inherited from one parent and *2 allele from the other
Genetic variants	A change or difference in the DNA sequence compared to a reference sequence. These changes can affect how genes function and may influence drug metabolism, efficacy, or risk of adverse effects
Genotype	A combination of alleles (versions of a gene) at a particular genetic location (locus), e.g., CC, CT
Germline mutation	Germline mutations occur in reproductive cells (egg or sperm) and can be passed from parent to child
Haplotype	A combination of genetic variants located close to each other on a single chromosome that tend to be inherited together
Pharmacogenes	Genes that influence how an individual responds to medications. These include genes involved in pharmacokinetics (absorption, distribution, metabolism, and elimination), pharmacodynamics (e.g., drug targets such as receptors), and immune response (e.g., human leukocyte antigen that may be associated with hypersensitivity reaction)
Phenotype	The observable trait and characteristic, e.g., rapid metabolizer
Single-nucleotide polymorphisms (SNPs)	A SNP occurs when there is a single nucleotide (A, T, C, or G) in the DNA sequence that is altered. A SNP can be synonymous or nonsynonymous. A synonymous SNP does not result in a change in amino acid sequence of a protein, while a nonsynonymous SNP results in a change in the amino acid sequence of a protein
Somatic mutation	Somatic mutations occur after conception in nonreproductive cells (not egg or sperm) and cannot be inherited by offspring
Star allele	The star allele nomenclature is commonly used to describe the allelic variation and haplotypes (combination of genetic variants on a single chromosome) of drug metabolizing enzymes, e.g., CYP2D6*1/*2

Table 2.2 Standardized PGx terms

Gene category	Phenotype term
Drug metabolizing enzymes (e.g., CYP2C19, DPYD, TPMT, UGT1A1)	Ultrarapid metabolizer Rapid metabolizer Normal metabolizer (formerly called extensive) Intermediate metabolizer Poor metabolizer
Transporter (e.g., ABCG2, SLCO1B1)	Increased function Normal function Decreased function Poor function
High-risk genotype status (e.g., HLA-B*57:01)	Positive Negative

Adapted from Ref. [2]

Genes are composed of DNA sequences that encode proteins. Genetic variants refer to differences in the DNA sequence compared to a reference sequence. A mutation is any change in the DNA sequence. These can be classified into germline and somatic mutations. Germline mutations occur in reproductive cells (egg or sperm) and can be inherited. Somatic mutations occur in nonreproductive cells after conception and are not passed to offspring.

A single nucleotide polymorphism (SNP) is a variation at a single nucleotide position in the DNA sequence. A synonymous SNP does not alter the amino acid sequence of the resulting protein. In contrast, a nonsynonymous SNP changes the amino acid sequence, potentially affecting the protein's structure, function, or stability of proteins. These changes may result in influencing disease susceptibility, drug response, or other traits.

Pharmacogenes are genes that influence drug response. PGx testing analyzes an individual's DNA—typically using buccal swabs, saliva, or blood samples—to identify relevant variants.

The star (*) allele nomenclature is widely used to describe variations in drug-metabolizing enzymes. The **1* allele typically represents the reference (wild-type, fully functional) haplotype (a combination of genetic variants on a single chromosome). Other alleles such as **2* or **3* define haplotypes carrying one more genetic variants [10]. In some cases, multiple copies of a gene may be present and are indicated with a suffix such as xN (e.g., "*1x3" for three copies of *1 allele). In PGx reporting, the combination of two star alleles inherited from each parent is referred to as a diplotype (e.g., CYP2C19*1/*1). The phenotype describes the observable characteristic (e.g., normal metabolizer). Phenoconversion refers to a mismatch between a patient's genotype-predicted phenotype and their actual observed phenotype.

PGx Testing and Reporting

PGx testing is considered a high-complexity clinical testing. Therefore, laboratories performing PGx analysis must meet the Clinical Laboratory Improvement Amendment (CLIA) or more stringent state-specific requirements [23]. Before using PGx results to guide clinical decisions, it is essential to confirm that the testing was conducted by a CLIA-certified laboratory. The Genetic Testing Registry can help identify testing centers to utilize in clinical practice when in-house testing is not offered [13].

The PGx laboratory will report a variant if one or more of the tested variants are found. If no variants are found, the result is reported as the reference (wild-type) allele—commonly designated as the **1* allele in the context of drug metabolizing enzymes [10].

It is important to note that there is no consistent way of translating genotype to phenotype results for pharmacogenes simply based on the specific number referenced in the star nomenclature convention. For example, *CYP2C19*2* allele codes for no function, and *CYP2D6*2* allele codes for normal function. Therefore, it is essential to check the genotype-to-phenotype translation for each pharmacogene. Fortunately, many PGx laboratories directly list phenotype information on the laboratory report, and many clinical guidelines (e.g., from the Clinical Pharmacogenetics Implementation Consortium, CPIC) list genotype-to-phenotype conversion tables to confirm the translation provided in the laboratory report. See the "real world look and feel" section below for guidance on interpreting these reports.

PGx Resources

ClinPGx is a centralized PGx resource launched in 2025 that integrates the work of three NIH-funded projects: Pharmacogenomics Knowledge Base (PharmGKB), Clinical Pharmacogenetics Implementation Consortium (CPIC), and Pharmacogenomics Clinical Annotation Tool (PharmCAT) [6]. It serves as a comprehensive PGx resource that (1) catalogs, curates, annotates, and disseminates PGx knowledge, (2) advances PGx implementation, (3) facilitates PGx integration, and (4) promotes PGx education.

PharmGKB is a PGx knowledge resource that helps clinicians and researchers understand how genetic variation influences drug response. It includes curated PGx literature, drug labels, and PGx guidelines from various countries for numerous drug–gene or variant pairs—including those without guidelines due to limited supporting data [8]. CPIC is one of the most widely accepted and endorsed sources of PGx clinical guidelines [5]. It is comprised an international consortium of staff and volunteers dedicated to facilitating the use of PGx in patient care. CPIC has developed peer-reviewed, evidence-based guidelines for over 100 medications. These guidelines are designed to help clinicians understand how to use available genetic results to optimize drug therapy, rather than to determine whether PGx testing should be ordered [16]. PharmCAT is a bioinformatics tool that analyzes genetic variants generated from sequencing or genotyping and translates them to gene-specific phenotypes and drug-prescribing recommendations [12].

The U.S. Food and Drug Administration (FDA) has an FDA Table of Pharmacogenomic Biomarkers in Drug Labeling, which lists drugs with PGx-related information in various sections of the label, such as boxed warning, dosage and administration, contraindications, and adverse reactions [21]. This table includes labels that mention PGx without offering dosing or drug selection guidance. It covers both somatic and germline PGx data.

In 2020, the FDA released a Table of Pharmacogenetic Associations highlighting germline drug-gene interactions with sufficient evidence [22]. These drug-gene associations are categorized into three sections: (1) those that support therapeutic management recommendations, (2) those indicating potential impact on safety or response, and (3) those affecting pharmacokinetic properties only. Inclusion in these tables does not imply the FDA recommends PGx testing prior to prescribing the medication unless the test is designated a companion diagnostic.

Activity Score System Used with Certain Pharmacogenes

Certain pharmacogenes are assigned an activity score (AS) to enhance the accuracy of phenotype prediction, especially when greater specificity is needed [7]. For example, the CYP2C9 intermediate metabolizer phenotypes are further classified into AS 1.5 and AS 1.0, with AS of 1.5 having a functional status closer to normal, while AS 1.0 exhibiting a relatively reduced function. Even these subtle differences can lead to distinct clinical implications and differing PGx recommendations for certain medications, such as piroxicam that is metabolized by CYP2C9 [19]. The AS is derived from a patient's genotype and currently applies to pharmacogenes such as *CYP2C9, CYP2D6,* and *DPYD*. For these genes, knowing a patient's AS is essential for determining the appropriate PGx-based guideline recommendations.

Phenoconversion

Phenoconversion can arise from drug–drug–gene interactions. For example, a strong CYP2CD6 inhibitor is prescribed to a patient who is a CYP2D6 normal metabolizer that can result in the patient behaving as a CYP2D6 poor metabolizer, potentially altering the metabolism of other CYP2D6 substrates the patient is taking. Other factors contributing to phenoconversion include aging, cancer, and inflammation [11]. CYP2D6 phenoconversion is most extensively studied as this enzyme metabolizes approximately 25% of commonly prescribed medications, including many antidepressants (see Chap. 4). An online CYP2D6 phenoconversion tool has been developed to support its clinical application [4]. Meanwhile, research continues to examine phenoconversion involving other pharmacogenes such as CYP2C19, particularly in the context of drug interactions [1, 17].

PGx Reports—Real-World Look and Feel

Since few institutions offer in-house PGx testing, most health care professionals rely on external PGx laboratories. While some electronic health record (EHRs) systems include PGx-specific fields—such as genotype, phenotype, and activity score—linked to clinical decision support tools, this level of integration is generally limited to select centers of excellence in the US. More commonly, PGx results from external laboratories are stored as scanned PDF reports in the media section of the EHR. In some cases, significant PGx findings may be manually entered as "allergies" or noted in clinical notes. This limits the utility of PGx to guide future prescribing decisions as PGx data is inadvertently buried or missed in a chart review. Additionally, most PGx reports include genotype and phenotype results with limited to no information on the associated medications.

Over the last two decades, PGx regulation in the US has shifted quite dramatically, resulting in vast variations in what is included in a PGx laboratory report. Recognizing that many prescribers may be relatively unfamiliar with PGx, it was once commonplace for PGx laboratories to include guidance for interpretation such as drug and dose selection recommendations. In 2019, the FDA issued a warning letter to one laboratory for "marketing certain genetic tests that have not been reviewed by the FDA for safety and effectiveness" with a further note that "these tests claim to predict patients' responses to specific medications based on genetic variants" [20]. Industry behaviors shifted following this announcement and warning letter to comply with this clarification of FDA's policy and requirements.

While many laboratories have reduced or discontinued medication-specific recommendations in their PGx reports, some laboratories still use a red/yellow/green color-coded system to indicate the magnitude of drug-gene interactions. This system, however, may be misinterpreted as "stop/caution/go"—akin to traffic signals—rather than its intended meaning of major, moderate, and minimal drug-gene interaction. A clear understanding of these color schemes is essential for appropriate clinical use of PGx results. For example, a major or moderate (red or yellow) drug–gene interaction may appear in a situation where the patient has been taking the affected medication for years with good therapeutic response. While this may seem counterintuitive, such a situation is not uncommon—especially in cases where PGx data would have been useful to guide drug initiation or dose titration, such as warfarin. The PGx data may, in hindsight, describe why a lower or higher than normal warfarin dose is necessary, potentially creating a more drawn-out titration timeline. However, once the warfarin dose is optimized and target INR is achieved, the therapy is effective despite not knowing PGx data at the time. In such cases, continuing the current regimen may be appropriate despite the red or yellow warning, particularly when other clinical data, such as INR, support therapeutic efficacy. This underscores that PGx is just one part of the broader clinical picture, and other factors must also be considered when making therapeutic decisions.

Practical PGx Considerations

Who and When to Test

Despite the evidence of PGx in reducing adverse events and improving outcomes, the routine use of PGx remains limited in the real-world setting. Several reasons have been cited for this delay in implementation, which include inadequate clinician education, slow test turnaround time, lack of integration into EHR, and poor reimbursement [24]. Compounding this is the lack of guidance on which patients to test and when to order a PGx test. Current PGx guidelines address how to use existing PGx data but not when to order the PGx test. The FDA provides guidance on genetically testing certain patients prior to therapy (e.g., *HLA-B*15:02* testing of genetically at-risk populations prior to initiating carbamazepine). However, no guidance exists on when or who to test, especially when PGx information could support dose modifications. This leaves the testing decision at the clinician's discretion.

Three PGx testing models have been proposed: (1) preemptive testing, (2) pretreatment testing, and (3) reactive testing [18]. Preemptive testing is similar to universal health screening where patients undergo PGx testing before any medication is prescribed. PGx results are available at the point of care, making this model ideal for managing chronic conditions and maximizing long-term utility of PGx results. However, it is rarely reimbursed. Pretreatment PGx testing is conducted when a drug with PGx considerations is about to be prescribed. This approach balances feasibility and clinical impact and is more likely to be reimbursed. Reactive testing occurs after treatment has begun, but in response to an adverse reaction or treatment failure. While this may inform future prescribing, it is less cost-effective and often too late to prevent harm from the original failed treatment. Although universal preemptive testing is theoretically ideal, real-world constraints favor a more targeted, diagnosis- or prescription-guided approach to PGx testing. There is increasing advocacy for stronger evidence-based guidance, integration into clinical workflows, and shared decision-making to fully realize the benefits of PGx in routine practice.

Which PGx Laboratory and Test to Use

The heterogeneity in PGx testing platforms presents a significant challenge to clinicians seeking to integrate precision medicine into routine care. The wide array of testing services offered by commercial laboratories, academic institutions, and healthcare systems creates confusion around test selection and results interpretation. Variability in test content, technologies, and reporting standards across platforms can lead to inconsistent genotype-to-phenotype translations, undermining confidence in the clinical utility of PGx testing. While issues of nomenclature consistency and genotype to phenotype translations have improved in recent years since the CPIC term standardization project, clinicians should be particularly mindful of older PGx reports that may benefit from a double check of genotype to phenotype conversions and terminology [2].

A critical concern is the variability in genetic variant coverage across different testing panels. While some platforms offer comprehensive multigene panels that include rare and population-specific variants, others may only assess a limited number of common alleles. This lack of standardization can result in different phenotype predictions for the same patient, depending on which test was used. For example, one panel might detect a CYP2D6 gene deletion or duplication that another test does not identify. Since the absence of a mutation often leads to the default interpretation of "normal metabolizer" phenotype, a missed variant can result in misclassifying the patient's metabolizer status to "normal metabolizer" from a significant variation such as "ultrarapid metabolizer." These discrepancies may significantly influence prescribing decisions, potentially affecting medication efficacy or increasing the risk of adverse events, leading to suboptimal treatment outcomes. Additionally, inconsistent nomenclature, reporting formats, and interpretation frameworks among testing vendors create further complexity. Clinicians—particularly those without specialized training in PGx—may struggle to interpret or act on conflicting reports. This challenge is compounded when results must be integrated into EHR or clinical decision support tools that rely on standardized data.

To address these issues, there is a pressing need for harmonization of PGx testing standards, including consensus on core gene-variant panels, uniform reporting practices, and standardized genotype-to-phenotype translation algorithms. Collaboration among professional organizations such as CPIC, the FDA, and clinical laboratories is essential to ensure the development of transparent, evidence-based, and clinically actionable tests.

In the interim, healthcare systems should consider establishing interdisciplinary teams—comprising laboratory directors, PGx-trained clinicians and pharmacists, and informaticians—who can guide PGx test selection, ensure accurate interpretation, and support PGx integration into clinical workflows. This team-based approach can enhance the clinical utility of PGx testing.

Sharing of PGx Data Across Different Care Settings

Effective integration of PGx data across different care settings (e.g., when patient transitions from hospital to nursing facility) is essential to maximize its clinical utility. However, interoperability challenges exist between health systems, which often limit data sharing. In many cases, the PGx report may need to be shared as faxed documents, with data not linked to decision support tools. This lack of integration can limit the clinical utility of PGx data, as it may not be readily available at the point of care to guide drug selection and dosing. Overcoming these barriers requires enhancements in data infrastructure to enable the seamless sharing of PGx information across diverse EHR systems and link them to clinical decision support tools.

PGx Cost and Reimbursement

PGx test coverage by various health plans can influence healthcare providers' ability to recommend PGx testing to patients. The cost of the test without insurance coverage can range from $200 to over $500 [15]. Within the US, some Medicare local coverage determinations (LCDs) may generally cover PGx testing in certain states when PGx testing is considered medically reasonable and necessary, and the patient will be prescribed PGx medications with CPIC level A or B designation (strong and moderate evidence, respectively). Private health plans offer variable coverage with a few covering multigene panels, though it should not be expected that preemptive screening where the patient will not immediately be prescribed a drug with PGx implications would be covered [18].

Reevaluation of PGx Results with Emerging Evidence

Although PGx results remain constant throughout a person's lifetime making retesting unnecessary, periodic re-evaluation of their clinical interpretation is important. As PGx knowledge evolves, the interpretation of specific genetic variants and the translation into phenotypes may change. For instance, the standardization of CYP2D6 genotype-to-phenotype translation by CPIC and Dutch Pharmacogenetics Working Group prompted many institutions to revise their PGx reports and reclassify affected patient phenotypes to align with the updated guidelines [3].

Reassessing PGx results in light of such updates helps to ensure more accurate phenotype classification and supports medication recommendations that reflect the latest scientific evidence. To facilitate this, healthcare systems may benefit from implementing mechanisms for regularly reviewing and updating PGx interpretations, particularly within EHR and clinical decision support tools, to improve the precision, relevance, and safety of pharmacotherapy over time. In situations where such infrastructure is not available, clinicians may find it helpful to check the date on a PGx report to identify whether key guideline updates have happened since the laboratory report was originally issued so that manual verification and updating can take place on a case-by-case basis.

PGx offers opportunity to further personalize medicine and should be interpreted alongside other clinical factors. In older adults, age-related physiologic changes as well as drug-drug-gene interactions, such as phenoconversion, can significantly influence medication metabolism and response. Integrating PGx insights with these considerations may help to optimize therapy and improve outcomes.

References

1. Abouir K, Exquis N, Gloor Y, Daali Y, Samer CF. Phenoconversion due to drug-drug interactions in CYP2C19 genotyped healthy volunteers. Clin Pharmacol Ther. 2024;116(4):1121–9. https://doi.org/10.1002/cpt.3378. Epub 2024 Jul 29
2. Caudle KE, Dunnenberger HM, Freimuth RR, et al. Standardizing terms for clinical pharmacogenetic test results: consensus terms from the Clinical Pharmacogenetics Implementation Consortium (CPIC). Genet Med. 2017;19(2):215–23. https://doi.org/10.1038/gim.2016.87. Epub 2016 Jul 21. PMID: 27441996; PMCID: PMC5253119
3. Caudle KE, Sangkuhl K, Whirl-Carrillo M, Swen JJ, Haidar CE, Klein TE, Gammal RS, Relling MV, Scott SA, Hertz DL, Guchelaar HJ, Gaedigk A. Standardizing CYP2D6 genotype to phenotype translation: consensus recommendations from the Clinical Pharmacogenetics Implementation Consortium and Dutch Pharmacogenetics Working Group. Clin Transl Sci. 2020;13(1):116–24. https://doi.org/10.1111/cts.12692. Epub 2019 Oct 24. PMID: 31647186; PMCID: PMC6951851.
4. Cicali EJ, Elchynski AL, Cook KJ, et al. How to integrate CYP2D6 phenoconversion into clinical pharmacogenetics: a tutorial. Clin Pharmacol Ther. 2021;110(3):677–87. https://doi.org/10.1002/cpt.2354. Epub 2021 Jul 28. PMID: 34231197; PMCID: PMC8404400.
5. Clinical Pharmacogenetics Implementation Consortium (CPIC). CPIC: optimizing drug therapy with pharmacogenetics. n.d. Retrieved September 1, 2025, from https://www.cpicpgx.org
6. ClinPGx. ClinPGx: clinical pharmacogenomics resource. 2025. Retrieved September 1, 2025, from https://www.clinpgx.org
7. Gaedigk A, Simon SD, Pearce RE, Bradford LD, Kennedy MJ, Leeder JS. The CYP2D6 activity score: translating genotype information into a qualitative measure of phenotype. Clin Pharmacol Ther. 2008;83(2):234–42. https://doi.org/10.1038/sj.clpt.6100406. Epub 2007 Oct 31
8. Gong L, Whirl-Carrillo M, Klein TE. PharmGKB, an integrated resource of pharmacogenomic knowledge. Curr Protoc. 2021;1(8):e226. https://doi.org/10.1002/cpz1.226. PMID: 34387941; PMCID: PMC8650697
9. Haga SB. The critical role of pharmacists in the clinical delivery of pharmacogenetics in the U.S. Pharmacy (Basel). 2023;11(5):144. https://doi.org/10.3390/pharmacy11050144. PMID: 37736916
10. Kalman LV, Agúndez J, Appell ML, et al. Pharmacogenetic allele nomenclature: international workgroup recommendations for test result reporting. Clin Pharmacol Ther. 2016;99(2):172–85. https://doi.org/10.1002/cpt.280. Epub 2015 Nov 20. PMID: 26479518; PMCID: PMC4724253.
11. Klomp SD, Manson ML, Guchelaar HJ, Swen JJ. Phenoconversion of cytochrome P450 metabolism: a systematic review. J Clin Med. 2020;9(9):2890. https://doi.org/10.3390/jcm9092890. PMID: 32906709; PMCID: PMC7565093
12. Li B, Sangkuhl K, Keat K, et al. How to run the pharmacogenomics clinical annotation tool (PharmCAT). Clin Pharmacol Ther. 2023;113(5):1036–47. https://doi.org/10.1002/cpt.2790. Epub 2022 Dec 23. PMID: 36350094; PMCID: PMC10121724

13. National Center for Biotechnology Information. Genetic Testing Registry (GTR). U.S. National Library of Medicine. n.d. Retrieved September 1, 2025, from https://www.ncbi.nlm.nih.gov/gtr/[1]. https://www.ncbi.nlm.nih.gov/gtr/tests/597430/
14. National Human Genomic Research Institute. Talking glossary of genomic and genetic terms https://www.genome.gov/genetics-glossary; Glossary of genomics terms. JAMA. 2013;309(14):1533–5. https://doi.org/10.1001/jama.2013.2950
15. Patel JN, Chaihorsky L, Dong OM, Lu CY, Moretz C, Reese E, Teeple W, Brown B, Rogers S. Medical policy determinations for pharmacogenetic tests among US health plans. Am J Manag Care. 2025;31(2):e47–55. https://doi.org/10.37765/ajmc.2025.89683.
16. Relling MV, Klein TE, Gammal RS, Whirl-Carrillo M, Hoffman JM, Caudle KE. The clinical pharmacogenetics implementation consortium: 10 years later. Clin Pharmacol Ther. 2020;107(1):171–5. https://doi.org/10.1002/cpt.1651. Epub 2019 Nov 5. PMID: 31562822; PMCID: PMC6925644
17. Scherf-Clavel M, Weber H, Unterecker S, et al. The relevance of integrating CYP2C19 phenoconversion effects into clinical pharmacogenetics. Pharmacopsychiatry. 2024;57(2):69–77. https://doi.org/10.1055/a-2248-6924. Epub 2024 Feb 14. PMID: 38354747; PMCID: PMC10948286
18. Stevenson JM, Smith DM, Tuteja S, Patel JN. Pharmacogenomics Global Research Network (PGRN) Publications Committee. Navigating pharmacogenomic testing in practice: who to test and when to test. Clin Pharmacol Ther. 2025;118(3):561–6. https://doi.org/10.1002/cpt.3704. Epub 2025 May 5. PMID: 40325943; PMCID: PMC12355021.
19. Theken KN, Lee CR, Gong L, et al. Clinical Pharmacogenetics Implementation Consortium Guideline (CPIC) for CYP2C9 and nonsteroidal anti-inflammatory drugs. Clin Pharmacol Ther. 2020;108(2):191–200. https://doi.org/10.1002/cpt.1830. Epub 2020 Apr 28. PMID: 32189324; PMCID: PMC8080882
20. U.S. Food and Drug Administration. FDA issues warning letter to genomics lab for illegally marketing genetic test claims to predict patients' responses. 2019. Retrieved September 1, 2025, from https://www.fda.gov/news-events/press-announcements/fda-issues-warning-letter-genomics-lab-illegally-marketing-genetic-test-claims-predict-patients
21. U.S. Food and Drug Administration. Table of pharmacogenomic biomarkers in drug labeling. n.d.-a. Retrieved September 1, 2025, from https://www.fda.gov/drugs/science-and-research-drugs/table-pharmacogenomic-biomarkers-drug-labeling
22. U.S. Food and Drug Administration. Table of pharmacogenetic associations. n.d.-b. Retrieved September 1, 2025, from https://www.fda.gov/medical-devices/precision-medicine/table-pharmacogenetic-associations
23. Vo TT, Bell GC, Owusu Obeng A, Hicks JK, Dunnenberger HM. Pharmacogenomics implementation: considerations for selecting a reference laboratory. Pharmacotherapy. 2017;37(9):1014–22. https://doi.org/10.1002/phar.1985. Epub 2017 Sep 3
24. Youssef E, Bhattacharya D, Sharma R, Wright DJ. A theory-informed systematic review of barriers and enablers to implementing multi-drug pharmacogenomic testing. J Pers Med. 2022;12(11):1821. https://doi.org/10.3390/jpm12111821. PMID: 36579514; PMCID: PMC9696651

Chapter 3
Integrating Pharmacogenomics into Geriatric Care: Challenges and Opportunities

Teresa DeLellis and Youssef Roman

Opportunities to Utilize Pharmacogenomics in Older Adults

The science of PGx is predicated on the knowledge of and selection and dosing based on known drug–gene interactions. Polypharmacy is widely prevalent in older adults, dramatically increasing the precedence of drug-gene interactions. Both drug–drug and drug–gene interactions increase the risk of adverse effects and treatment failure in older adults, and when they coincide, drug–drug–gene interactions could occur [2, 13]. In one study of a Program for All-inclusive Care for the Elderly (PACE) population, 125 drug-drug-gene interactions were identified across 100 participants [1].

Challenges in Pharmacogenomics in Older Adults

Pharmacogenomic (PGx) testing has gained support in recent literature for its role in reducing medication treatment failures and adverse reactions. However, much of this evidence is primarily derived from studies conducted in middle-aged populations without the multicomplexity observed in geriatric care. As a result, applying PGx insights in the field of clinical geriatrics presents unique challenges.

T. DeLellis (✉)
School of Pharmacy, College of Health Professions, Nursing and Pharmacy, Manchester University, Fort Wayne, IN, USA
e-mail: TMDeLellis@manchester.edu

Y. Roman
L.S. Skaggs College of Pharmacy, Idaho State University, Pocatello, ID, USA
e-mail: youssefroman@isu.edu

T. DeLellis et al. (eds.), *Applying Pharmacogenomics to Optimize Geriatric Care*, https://doi.org/10.1007/978-3-032-10653-7_3

Older adults are often underrepresented in clinical trials, including those focused on PGx. Consequently, much of the available evidence guiding treatment decisions comes from studies conducted in younger or middle-aged populations. This underrepresentation obscures how age-related factors—comorbidities, polypharmacy, and physiological changes—affect treatment outcomes. The problem is further compounded in genetic-based studies, where findings are frequently extrapolated from younger cohorts, overlooking age-related genetic, metabolic, and physiological differences. This creates significant gaps in evidence-based care for older adults, leaving clinicians to navigate critical decisions with limited, and sometimes inappropriate data.

Pharmacogenomics in older adults is further complicated by age-specific factors that reduce its predictive accuracy and clinical utility. Despite being the population most likely to use multiple medications, older adults are often underrepresented in PGx trials [6, 9]. Age-related physiological changes in pharmacokinetics (e.g., reduced hepatic metabolism, impaired renal clearance, and altered drug distribution) and pharmacodynamics (e.g., altered receptor sensitivity) can modify or obscure the impact of genetic variants observed in younger populations [3, 4, 10]. In addition, multimorbidity, polypharmacy, and inflammation contribute to "phenoconversion," where an individual's observed drug response deviates from their genotype-predicted response, further reducing the consistency of effect sizes across age groups [8, 10].

Effects of drug-gene interactions have been documented to show variability based on interactions with aging physiology. For example, older adults who are *CYP2D6* poor metabolizers have an 18-fold higher plasma concentration of venlafaxine compared to younger adults with the same phenotype [15]. The overlay of aging on top of PGx information is not limited to objective lab values like drug concentrations, but is also demonstrated in more subjective drug response characteristics. For example, another study retrospectively analyzed the phenotype of 173 patients aged 70 years and older on an average of 10 medications or more [14]. Results demonstrated significantly higher rates of orthostasis, a common geriatric syndrome that complicates optimization of mortality-reducing medication management in patients with reduced or absent *CYP2D6* activity. The study also revealed that equivalent doses were utilized across all phenotypes in the study, suggesting that application of PGx data could have reduced the incidence of orthostasis.

As another example, aging can result in downregulation of beta-1 receptors, decreasing response to beta-blockers in older adults. Metoprolol is metabolized by *CYP2D6*, and those with *CYP2D6* ultra-rapid metabolizer status need higher doses to achieve therapeutic effects. This effect would be compounded in older adults compared to younger adults due to the pharmacodynamic effects of aging that blunt metoprolol effects. However, clinicians should also be mindful of cautious dose escalation to avoid orthostasis [13]. Therefore, the complexity of layering PGx decisions on top of geriatric pharmacotherapy management requires careful consideration by clinicians who simultaneously balance several aspects of risk/benefit for their patients.

Pharmacogenomic testing may help simplify other clinical decisions. Discovering phenotypes that support likely ineffectiveness or increased risk of adverse effects even at the lowest doses could help overcome clinical inertia toward optimizing medication regimens in older adults, particularly in cases of geriatric syndromes and when deprescribing is indicated. For example, musculoskeletal discomfort is widespread in older adults, with an equally widespread list of possible causes. A low-intensity statin being utilized for primary prevention in an older adult with low to moderate cardiovascular risk might quickly be dismissed as a cause and not considered for deprescribing. However, a phenotype showing reduced SLCO1B1 activity in that patient could create significant additional concern (as will be discussed more specifically in upcoming chapters), and support deprescribing [13].

Pharmacogenomics and Initial Dosing

Selecting the appropriate starting dose of medications in older adults is inherently complex. Physiologic changes of aging, multimorbidity, polypharmacy, and the unpredictability of drug–drug and drug–disease interactions create an entangled therapeutic balance. In many cases, prescribing guidelines recommend a range for the initial dose rather than a fixed dose, reflecting the clinical uncertainty around interindividual variability in drug response. For older adults, this flexibility presents both an opportunity and a challenge: the choice of whether to start at the low, mid, or upper end of the recommended range can influence therapeutic success, tolerability, and safety.

Pharmacogenomic information offers a novel, evidence-based dimension to inform these initial dosing decisions. Unlike traditional clinical factors such as age, renal function, or comorbidities, PGx data provides insight into the individual's capacity to metabolize or transport drugs before therapy even begins. For example, knowing that an older adult carries a poor metabolizer genotype for *CYP2D6* can justify starting at the lowest end of a dosing range for a drug heavily dependent on this pathway. Conversely, identifying an ultra-rapid metabolizer status may support the rationale for starting at the higher end of the range to avoid subtherapeutic exposure.

In this way, PGx information does not replace but rather complements geriatric principles of "start low and go slow." It allows the "low" or "high" end of a dosing spectrum to be anchored in biologically meaningful evidence rather than left solely to clinical judgment or trial-and-error. This can be particularly valuable in reducing adverse events, already more common in older adults, while also minimizing delays in achieving therapeutic benefit.

Ultimately, the integration of PGx data as an added level of evidence could transform dosing ranges from broad, population-based recommendations into more precise, individualized starting points. While additional research is needed to solidify emerging evidence of cost-effectiveness and real-world outcomes in geriatric populations, the potential is clear: PGx-guided initial dosing could enhance both safety and efficacy in one of the most vulnerable patient groups.

Pharmacogenomics and Risk Stratification

The AGS Beers Criteria® highlights medications that are potentially inappropriate in older adults, often categorized as "use with caution" because risk is elevated in certain situations, but not uniformly across all patients. One of the persistent challenges in applying the AGS Beers Criteria® is its broad scope: clinicians are left with population-level warnings but little individualized guidance on who, specifically, might tolerate these medications safely.

Pharmacogenomics (PGx) provides an added layer of evidence that can refine these risk assessments. By integrating PGx data, clinicians can move beyond a one-size-fits-all "caution" label to a more personalized risk–benefit framework:

- **Medication metabolism and exposure**: Genetic variants related to metabolizing enzymes and transporters can influence medication levels. Several medications that require "use with caution," such as certain tricyclic antidepressants, SSRIs, opioids, or antiplatelets, have well-characterized PGx variability in CYP2D6, CYP2C19, or other metabolizing enzymes. For medications that are pharmacologically active, patients who are normal or rapid metabolizers may have a lower risk of medication accumulation and related adverse effects compared to poor metabolizers.
- **Pharmacodynamic targets**: Genetic variants that affect medication targets—such as receptors or transporters—can alter how a patient responds to treatment. Older adults who carry genetic alleles associated with reduced drug response may face less "harm" at standard doses compared to those with increased sensitivity.
- **Layered decision-making**: Rather than treating the AGS Beers Criteria® "caution" as a binary yes/no, clinicians can use PGx results to stratify patients:
  - High-risk PGx phenotype + "use with caution" medication = may yield a stronger reason to avoid or seek alternatives.
  - Low-risk PGx phenotype + "use with caution" medication = suggests a case where benefits may outweigh risks, especially if therapeutic alternatives are limited.
- **Reducing therapeutic denial**: One unintended consequence of the AGS Beers Criteria® is over-avoidance of useful medications, which can impair quality of life (e.g., avoiding effective antidepressants or pain medications). PGx can help identify which patients are the lowest risk candidates for a given therapy, preserving treatment options while maintaining patient safety.

Ultimately, PGx does not replace the AGS Beers Criteria® criteria but adds precision, helping to identify subsets of older adults for whom "caution" can be reinterpreted, considering their genetic risk. This integration could support a shift toward more individualized prescribing in geriatrics, where both over- and under-treatment remain pressing concerns. One example application is in reducing falls in older adults [7]. A wide array of falls risk-inducing drugs (FRIDs) have PGx guidance available. While many of these are also designated as potentially inappropriate medications (PIMs) in the AGS Beers Criteria®, use is still common and, in some situations, justified. Harnessing PGx would be a great use in situations where PIMs or FRIDs may be a necessary risk, adding further risk and benefit data based on a person's phenotypes to optimize treatment selection and dosing strategies [7].

When PGx Testing May Not Add Value

Despite the potential clinical utility of PGx testing in older adults, it may not be appropriate in certain clinical scenarios. For older patients with limited life expectancy and no anticipated changes to their medication regimen, PGx testing is unlikely to offer meaningful benefit. Similarly, if a patient is not currently taking, nor expected to take, medications with known PGx relevance, testing may not be warranted.

Emerging Evidence in Older Adults

Despite all these compounded complexities and the theoretical nature of many of the proposed benefits, applying PGx information to clinical decision making in older adults and other polypharmacy populations has been demonstrated to improve clinical and cost-related outcomes. Pharmacogenomic testing can be particularly valuable in older adults with polypharmacy, as the use of multiple medications increases the risk of adverse drug events and drug-gene interactions. Prioritizing genetic variants that affect high-risk medications—such as anticoagulants, antiplatelets, antidepressants, and opioids—can help guide drug selection and dosing of these drugs. Pharmacogenomic results may also support deprescribing efforts or guide the selection of safer alternatives tailored to individual genetic profiles. Vast improvements have been demonstrated in symptom resolution, reduced adverse effects, emergency department visits, re-hospitalizations, and net cost-savings after factoring in the cost of testing [5, 11]. There have been some mixed results reported; the studies with the most dramatic and consistent positive outcomes included consideration of drug-drug-gene interactions, rather than simply drug-gene interactions alone.

Importantly, PGx presents only one piece of the puzzle and must be layered with nongenetic factors such as physiological changes with age and multicomplexity interactions that impact the ultimate medication response [3, 10]. Finally, inequities in ancestry representation intersect with age under-sampling, limiting the generalizability of findings [12]. Together, these factors make pharmacogenomic implementation in geriatrics more complex and call for age- and context-specific evidence.

References

1. Bain KT, Matos A, Knowlton CH, McGAin D. Genetic variants and interactions from a pharmacist-led pharmacogenomics service for PACE. Pharmacogenomics. 2019;20(10):709–18. https://doi.org/10.2217/pgs-2019-0047.
2. Bozina N, Kirhmajher MV, Simicevic L, et al. Use of pharmacogenomics in elderly patients treated for cardiovascular diseases. Croat Med J. 2020;61:147–58. https://doi.org/10.3325/cmj.2020.61.147.
3. Drenth-van Maanen AC, et al. Prescribing medicines to older people. How to consider the impact of ageing on human organ and body functions. Br J Clin Pharmacol. 2020;86(10):1921–30. https://doi.org/10.1111/bcp.14094.
4. Ferrara N, et al. β-adrenergic receptor responsiveness in aging heart and clinical implications. Front Physiol. 2014;4 396 https://doi.org/10.3389/fphys.2013.00396.
5. Hayashi M, Hamdy DA, Mahmoud SH. Applications for pharmacogenomics in pharmacy practice: a scoping review. Res Soc Admin Pharm. 2022;18:3094–118. https://doi.org/10.1016/j.sapharm.2021.08.009.
6. Herrera AP, et al. Disparate inclusion of older adults in clinical trials: priorities and opportunities for policy and practice change. Am J Public Health. 2010;100(Suppl 1):S105–12. https://doi.org/10.2105/AJPH.2009.162982.
7. Just KS, Schneider KL, Schurig M, et al. Falls: the adverse drug reaction of the elderly and the impact of pharmacogenetics. Pharmacogenomics. 2017;18(13):1281–97. https://doi.org/10.2217/pgs-2017-0018.
8. Klomp SD, et al. Phenoconversion of cytochrome P450 metabolism: a systematic review. J Clin Med. 2020;9(9):2890. https://doi.org/10.3390/jcm9092890.
9. Lau SWJ, Huang Y, Hsieh J, et al. Participation of older adults in clinical trials for new drug applications and biologics license applications from 2010 through 2019. JAMA Netw Open. 2022;5(10):e2236149. https://doi.org/10.1001/jamanetworkopen.2022.36149.
10. Lenoir C, et al. Influence of inflammation on cytochromes P450 activity in adults: a systematic review of the literature. Front Pharmacol. 2021;12:733935. https://doi.org/10.3389/fphar.2021.733935.
11. O'Shea J, Ledwidge M, Gallagher J, Keenan C, Ryan C. Pharmacogenetic interventions to improve outcomes in patients with multimorbidity or prescribed polypharmacy: a systematic review. Pharmacogenomics J. 2022;22:89–99. https://doi.org/10.1038/s41397-021-00260-6.
12. Roman Y. Bridging the United States population diversity gaps in clinical research: roadmap to precision health and reducing health disparities. Personal Med. 2025;22(3):193–203. https://doi.org/10.1080/17410541.2025.2504329.
13. Roman YM, Dixon DL, Salgado TM, et al. Challenges in pharmacotherapy for older adults: a framework for pharmacogenomics implementation. Pharmacogenomics. 2020;21(9):627–35. https://doi.org/10.2217/pgs-2019-0198.
14. Roskaug R, Wyller TB, Straand J, Kersten H, Molden E. Prescribed doses of FYP2D6-metabolized drugs and hemodynamic responses in relation to CYP2D6 genotype among older patients exposed to polypharmacy. Drugs Aging. 2020;37:425–33. https://doi.org/10.1007/s40266-020-00763-0.
15. Schlender JF, Vozmediano V, Golden AG, et al. Current strategies to streamline pharmacotherapy for older adults. Eur J Pharmaceut Sci. 2018;111:432–42. https://doi.org/10.1016/j.ejps.2017.10.015.

Part II
Application

Chapter 4
Pharmacogenomics and Gastrointestinal Care in Older Adults

Stephanie Sibicky, Teresa DeLellis, Yee Ming Lee, David Bright, and Youssef Roman

Introduction

Aging-related changes to the gastrointestinal (GI) tract have the potential to impact the effect of medications in older adults, alone or in combination with pharmacogenomic considerations. From a physiologic perspective, aging impacts the entire GI system, causing a litany of alterations to account for when administering medications and can compound adverse effects [4]. For example, a reduction in saliva production and quality can lead to xerostomia, which is also a common side effect associated with anticholinergic medications. The reduction in stomach acid and increase in pH (i.e., achlorhydria) can lead to ulcers, gastroesophageal reflux disease (GERD), and increase the risk for infection. When thinking about the pharmacokinetics of medications and the impact of aging on the GI tract, absorption of medications remains mostly unchanged; however, distribution of medications may

S. Sibicky (✉)
Clinical Professor and Assistant Dean for Academic Affairs and Assessment, Bouvé College of Health Sciences, Northeastern University, Boston, MA, USA
e-mail: s.sibicky@northeastern.edu

T. DeLellis
School of Pharmacy, College of Health Professions, Nursing and Pharmacy, Manchester University, Fort Wayne, IN, USA

Y. M. Lee
Associate Professor of Clinical Pharmacy, Skaggs School of Pharmacy and Pharmaceutical Sciences, University of Colorado, Aurora, CO, USA

D. Bright
Professor of Pharmaceutical Sciences, College of Pharmacy, Ferris State University, Big Rapids, MI, USA

Y. Roman
L.S. Skaggs College of Pharmacy, Idaho State University, Pocatello, ID, USA

T. DeLellis et al. (eds.), *Applying Pharmacogenomics to Optimize Geriatric Care*, https://doi.org/10.1007/978-3-032-10653-7_4

decrease with a reduction in serum albumin (especially for protein-bound drugs). Metabolism, likely affected by a reduction in liver size and hepatic blood flow, can also alter the extraction ratio of certain medications; if a medication is highly dependent on hepatic blood flow, dose adjustments may be necessary (e.g., amitriptyline). Medications with a low hepatic extraction may be more dependent on enzyme activity compounded by the potential for less enzyme content in the liver [8]. There is emerging evidence regarding changes to gastric emptying and shifts in gut microbiota, although the effect on medication pharmacokinetics and pharmacodynamics is unclear [2].

While dosing changes may need to be considered due to the age-related changes to the GI tract, pharmacogenomics may also prove useful in managing medication use in this population. Mostly, pharmacogenomics (PGx) panels ordered for other conditions—such as psychiatric medications—can be utilized for GI medications as well, since many of the same enzymes will be included in both panels. This information has longevity; a patient's pharmacogenomics do not "age" and can be used in perpetuity for that patient.

Proton Pump Inhibitors and the Role of CYP2C19

Proton pump inhibitors (PPIs) have evolved across several generations since their introduction in the late 1980s. The first-generation PPIs (e.g., omeprazole, lansoprazole) were among the earliest acid-suppressing agents that irreversibly inhibited the H^+/K^+ ATPase in gastric parietal cells. These drugs are extensively metabolized by the CYP2C19 enzyme, leading to significant interindividual variability in plasma concentrations and therapeutic outcomes. As pharmacogenomic understanding grew, newer agents such as esomeprazole (a stereoisomer of omeprazole) and dexlansoprazole (a modified-release formulation of lansoprazole) were developed to provide more consistent acid suppression with less variability. However, despite formulation improvements, many newer PPIs still rely on CYP2C19 for metabolism to varying degrees [13].

Current CPIC guidelines emphasize the clinical relevance of CYP2C19 polymorphisms in determining PPI efficacy and risk (see Table 4.1) [9]. For example, CYP2C19 ultrarapid metabolizers (UMs) often under-respond to standard doses due to rapid drug clearance, warranting a higher or more frequent dose. In contrast, CYP2C19 poor metabolizers (PMs) may have supra-therapeutic drug concentrations, increasing the risk for long-term adverse effects such as nutrient malabsorption, vitamin deficiency, or kidney injury. CPIC provides level A recommendations for PPIs like omeprazole, lansoprazole, and pantoprazole, while the guidance for esomeprazole and rabeprazole remains at level C due to insufficient evidence linking CYP2C19 genotype with clinical outcomes.

Table 4.1 CPIC guideline for proton pump inhibitors (omeprazole, lansoprazole, pantoprazole, and dexlansoprazole) based on CYP2C19 phenotype [9]

<table>
<tr><th>CYP2C19 phenotype</th><th>CPIC recommendations[a]</th></tr>
<tr><td>CYP2C19 ultrarapid metabolizer</td><td>Increase starting daily dose by 100%. Daily dose may be given in divided doses. Monitor for efficacy.</td></tr>
<tr><td>CYP2C19 rapid metabolizer</td><td rowspan="2">Initiate standard starting daily dose. Consider increasing dose by 50–100% for the treatment of Helicobacter pylori infection and erosive esophagitis. Daily dose may be given in divided doses. Monitor for efficacy.</td></tr>
<tr><td>CYP2C19 normal metabolizer</td></tr>
<tr><td>CYP2C19 intermediate metabolizer</td><td rowspan="2">Initiate standard starting daily dose. For chronic therapy (>12 weeks) and if efficacy is achieved, consider 50% reduction in daily dose and monitor for continued efficacy.</td></tr>
<tr><td>CYP2C19 poor metabolizer</td></tr>
</table>

[a]Alternative PPI options that are less affected by CYP2C19 include esomeprazole and rabeprazole

Ondansetron and the Role of CYP2D6

Ondansetron, a 5-HT_3 receptor antagonist used for chemotherapy and surgery-induced nausea and vomiting, is primarily metabolized by CYP2D6. Patients with ultrarapid CYP2D6 activity may rapidly clear ondansetron, leading to subtherapeutic drug levels and inadequate control of nausea. CPIC guidelines recommend avoiding ondansetron in such individuals and suggest using alternative agents like granisetron, which are not dependent on CYP2D6 metabolism [3]. Unlike PPIs, where dose adjustments may be feasible, ondansetron does not have well-supported higher-dose strategies, so genotype-guided drug selection is preferred to dose changes.

Case 1

A 71-year-old male presents to his primary care provider's (PCP) office with complaints of a near-constant upset stomach unless he's recently eaten something "soothing." He also complains of acid reflux symptoms. He reports no recent medication changes except for taking "some OTC ibuprofen" following a pickleball injury 3 weeks ago (right shoulder). His primary care physician ordered PGx testing 8 years ago when he was considering an SSRI for depression following the death of the patient's spouse; however, the patient eventually opted for counseling instead of medications. The limited PGx panel that was originally intended for mental health medication guidance is still available in the medical record.

Medical History:

Dyslipidemia (18 years)
Hypertension (24 years)

Allergies:

No known drug allergies

Pertinent Vitals and Labs:

CrCl 70–75 mL/min

Pharmacogenomic Results:

CYP2D6: normal metabolizer
CYP2C9: poor metabolizer
CYP2C19: intermediate metabolizer

Current Medications:

Tums as needed
Rosuvastatin 20 mg by mouth QHS
Lisinopril/HCTZ 10 mg/25 mg by mouth daily
Ibuprofen 200 mg, two tabs by mouth Q6H PRN (typically taking 6–8 tabs daily)

Clinical (Nongenetic) Considerations

As a natural course of aging, gastric acid secretion changes can occur. The most salient considerations in this case include a decrease in prostaglandin production leading to impaired gastric cytoprotection, coinciding with a decrease in gastric acid production that can interfere with micronutrient absorption [7, 10]. Nonsteroidal anti-inflammatory drugs (NSAIDs) pose many health risks to older adults and are likely the culprit of his gastric complaints through further inhibition of prostaglandin production in the stomach. Additional NSAID concerns include increased risk of acute kidney injury in combination with lisinopril, worsening hypertension potentially leading to further antihypertensive medication use, and increasing cardiovascular risk. The AGS Beers Criteria® recommend avoiding chronic use of NSAIDs in older adults unless a gastroprotective agent such as a proton pump inhibitor (PPI) can be taken in combination [1]. The prescribing cascade effect of adding a PPI to preserve NSAIDs as a therapeutic option includes its own risks. PPI use has been associated with *Clostridium difficile* infections, pneumonia, GI malignancies, bone loss, cognitive impairment, vitamin deficiencies, and fractures, and thus is also recommended to avoid for long-term use (>8 weeks). This patient does not present with complex multimorbidity, and the pathologies of his medical conditions do not complicate the case beyond the medication interactions already discussed.

Pharmacogenomics Considerations

The patient's PGx results lend further credence to the clinical geriatrics concerns in this case. Ibuprofen is metabolized by CYP2C9 (see Chap. 4, Table 4.1). The patient has a CYP2C9 poor metabolizer phenotype indicating that a reduced metabolism will cause the ibuprofen AUC to be increased. Excessive exposure to ibuprofen via this drug-gene interaction enhances the risk of adverse effects associated with NSAID use [9]. The aforementioned warning against chronic NSAID use becomes heightened with respect to ibuprofen given the PGx data observed. However, in this short-term NSAID situation, and assuming reasonable renal function (CrCl >30 mL/min), the patient should be advised to utilize a lower dose of ibuprofen or select an NSAID such as naproxen (no more than 600 mg/day) that is not primarily metabolized by CYP2C9, given the PGx data observed. With the GI adverse effects described, it is unlikely that an NSAID will be continued. Alternatively, acetaminophen (<3 g/day) could be a therapeutic option and does not require dose adjustments based on the PGx data shown.

This patient will likely require a short-term PPI to assist in recovery from this suspected NSAID-induced gastritis. Omeprazole is a commonly used PPI because it is available over the counter and at a relatively low cost. Because omeprazole is metabolized by CYP2C19 and the patient is a CYP2C19 IM, a PGx report may indicate caution for the drug-gene combination. However, CPIC guidelines do not recommend altering *initial* dosing for omeprazole (Table 4.1) [9]. Instead, the recommendation is to reduce the dose after 12 weeks if chronic use is planned. This is an example of how the "stoplight" based visual guidance provided with PGx results reports can be misleading, and deeper investigation into actual guidance is warranted. Because PPI use should be limited to fewer than 8 weeks in older adults already, this patient should be able to use either normal-dose omeprazole or another PPI that is not metabolized by 2C19 unless prolonged therapy is deemed necessary through shared decision-making. If this patient had additional risk factors, such as infections, GI malignancies, low bone-mineral density, fractures, or showed signs of aging leading to lower baseline gastric acid production, they likely would not closely resemble patients in the studies that formulated the CPIC recommendations. Therefore, from a clinical geriatric standpoint, it would still be reasonable to start at a lower dose of omeprazole (i.e., 20 mg) or consider alternative acid-suppression therapy, such as famotidine 20 mg daily.

There are other PGx considerations for this patient based on his phenotype and statin use. Refer to Chap. 6 for an in-depth discussion on this topic.

Overall Recommendation

- This patient's current NSAID use should be discontinued in favor of nonsystemic NSAID multimodal pain strategies as discussed in Chap. 4.

- Future short-term systemic NSAID use, if indicated, should include a lower ibuprofen dose (200 mg) or switch to naproxen with a maximum of 600 mg per day accompanied with a PPI.
- Use of a PPI to help resolve current symptoms or for future gastroprotection with systemic NSAIDs can be selected and dosed according to package labeling unless risk factors for PPI sequelae or signs of advanced aging are present. In that case, if omeprazole is chosen, it may be reasonable to initiate a 20 mg starting dose or alternative acid-suppressing therapy.

What PGx Changed

- While an NSAID-induced gastritis likely didn't require any confirmation via PGx data, the PGx information provided useful guidance for how to more safely use different NSAID dosing and/or NSAID selection for future NSAID needs.
- PGx guidance for PPIs did not definitively change recommendations for appropriate use of PPIs in this patient. It confirmed that the lower end of a standard starting dose would be safest.

Case 2

A 70-year-old female with a past medical history significant for hypertension, hyperlipidemia, and irritable bowel syndrome with diarrhea (IBS-D) presents to her PCP for a follow-up after a recent viral gastroenteritis. She has some nausea symptoms with persistent mild diarrhea (one to two times a week) in addition to feeling increasingly anxious (increased worry about health status, heart racing). Her PCP ruled out infective causes and is considering a low-dose tricyclic antidepressant for her IBS-D symptoms and anxiety, and an antiemetic for her residual nausea.

Medical History:

Hypertension (25 years)
Hyperlipidemia (22 years)
IBS-D (30 years)
Coronary artery disease (CAD; 5 years)
Anxiety (30 years)

Allergies:

No known allergies

Pertinent Vitals and Labs:

Clinic BP 121/80, HR 75. CrCl = 65 mL/min
All other results within normal limits.

Pharmacogenomic Results:

*CYP2D6 *4/*5*: Poor metabolizer
*CYP2C19 *1/*1*: Normal metabolizer

Current Medications:

Amlodipine 5 mg by mouth daily
Metoprolol tartrate 25 mg by mouth BID
Paroxetine 20 mg by mouth daily
Atorvastatin 80 mg by mouth qHS
Aspirin 81 mg by mouth daily

Clinical (Nongenetic) Considerations

Nausea is one of the most common symptoms in ill patients and additionally presents as a frequent atypical symptom of almost any illness in frail older adults [6]. Of the many potential noninfectious causes, in this case, the patient's anxiety is the most likely contributor, particularly considering the timing with the IBS-D flare. Most antiemetic medications carry unfavorable risk-benefit profiles in older patients. Thus, in most cases, they should be reserved for when addressing causative factors resulting in insufficient relief or when doing so will take an extended amount of time, and short-term relief is needed. If this case is truly influenced by uncontrolled anxiety, the latter situation is likely. The patient should also be assessed for symptoms of hypotension, which can also induce nausea. Of her current medications, aspirin is the most likely to cause GI upset, but it is much less common at a low dose and is more likely to present as stomach pain or heartburn symptoms. Changing to an enteric formulation can help ensure this is not a contributing factor.

Standard approaches to treatment for anxiety in older people should be undertaken and are described in detail in Chap. 7. If short-term medications for nausea are desired to relieve symptoms in the meantime, metoclopramide, phenothiazines (e.g., prochloperazine), and ondansetron are commonly prescribed options to consider. Metoclopramide and phenothiazines are both potentially inappropriate medications per the AGS Beers Criteria [1]. Metoclopramide increases the risk of drug-induced Parkinson's disease, and exacerbation of anxiety and depression may be heightened in older adults. Phenothiazines have anticholinergic properties, which may help with IBS-D symptoms, but have many other risks in older adults including cognitive impairment, dry mouth affecting nutritional status, and falls risk. Serotonin receptor antagonists, such as ondansetron, are generally better tolerated. One of the most common adverse effects associated with ondansetron is constipation, which could be helpful in resolving this patient's diarrhea, and may have an additive effect with potential constipating effects of amlodipine. While amlodipine is not as likely to cause constipation as the non-dihydropyridine (N-DHP) calcium channel blockers, the likelihood increases as aging progresses and multimorbidity and polypharmacy advances.

Treatment of IBS-D is especially complex in older adults, particularly considering the anticholinergic effects of many of the treatment options with the best evidence per the American Gastroenterology Association (AGA) guidelines (tricyclic antidepressants (TCAs), antispasmotics) [11]. TCAs can also prolong the QT interval, as several of this patient's antiemetic treatment options, including ondansetron, do as well [1, 11]. While "as needed" use may be justified in some patients with severe symptoms, anticholinergic effects would be compounded when used in combination with the patient's paroxetine. Alternative SSRIs will need to be chosen carefully, prioritizing those with the lowest risk for diarrhea-related adverse effects. Beta-blockers can also cause or exacerbate diarrhea in older adults. In this case, changing the patient's antihypertensive regimen to a N-DHP calcium channel blocker would consolidate therapy for both hypertension and any anginal symptoms associated with their CAD, while assisting with diarrhea symptoms by virtue of their stronger constipating potential. However, there are other cascading effects with this approach that could complicate benefits. It is possible to transition IBS-D to IBS with constipation (IBS-C) through over-treatment, and N-DHPs are much stronger CYP3A4 and CYP3A5 inhibitors, so atorvastatin may need to be adjusted as well. Other options suggested by the AGA for IBS-D symptoms include loperamide, eluxadoline, and rifaximin. Loperamide and eluxadoline both have mixed opioid receptor effects and could exacerbate the patient's nausea. Rifaximin can also cause nausea, but the incidence is much lower when used in patients with IBS-D for the indicated duration of 14 days [14].

Pharmacogenomic Considerations

Of the antiemetic treatment options for this patient, PGx guidance does not currently exist for prochlorperazine. Based on their CYP2D6 poor metabolizer phenotype, the patient is at risk for accumulation of metoclopramide. The FDA label thus recommends starting at a lower than standard dose of 5 mg up to four times daily as needed (Table 4.2) [12]. Ondansetron can also accumulate in patients with CYP2D6 poor metabolizer phenotype, but based on clinical trials, current CPIC guidelines do not recommend altering the standard starting dose (Table 4.3) [3]. In light of the current clinical elements and contemporary guidelines, a dose adjustment for ondansetron is therefore not warranted despite the patient being CYP2D6 PM.

Of the IBS-D treatment options, TCAs have the most PGx guidance available [5]. Currently, PGx guidance does not exist for antispasmodics, loperamide, eluxadoline, or rifaximin. All TCAs (amitriptyline, nortriptyline, imipramine, desipramine) are metabolized via CYP2D6 and would be at risk of accumulation in this patient with CYP2D6 poor metabolizer phenotype. Some TCAs offer additional metabolism via CYP2C19, but this patient is a CYP2C19 NM and therefore no

Table 4.2 FDA pharmacogenomic information for metoclopramide and CYP2D6 [12]

CYP2D6 phenotype	CPIC recommendations
CYP2D6 poor metabolizer	For gastroesophageal reflux disease, use 5 mg by mouth four times a day or 10 mg by mouth three times a day (max 30 mg per day). For diabetic gastroparesis, use 5 mg by mouth four times a day (max 20 mg per day).

Table 4.3 CPIC guideline for ondansetron and CYP2D6 [3]

CYP2D6 phenotype	CPIC recommendations
CYP2D6 ultrarapid metabolizer	Consider an alternative antiemetic (e.g., granisetron) due to increased ondansetron metabolism with the risk of decreased drug response. Note: Dolasetron, palonosetron, and ramosetron are metabolized by CYP2D6, but there is limited data to dose based on CYP2D6 results; hence, no recommendations are currently available.

PGx-driven adjustments are necessary for the CYP2C19 component of the metabolism. Because the CPIC guideline recommendations for TCAs are based on data evaluating higher TCA doses used to treat depression, and the lower TCA doses (e.g., amitriptyline 10–25 mg at bedtime) used in IBS-D are less likely to cause adverse effects, dose adjustments are not routinely recommended. In this case, pharmacogenomic recommendations should be indication specific. As noted in the ondansetron discussion, individual patient characteristics warrant additional consideration, particularly when considering the heightened risks of anticholinergic use in older adults. If prescribed, the lowest effective dose should be used, and it should be titrated slowly with close monitoring of anticholinergic, cardiac, and neurological adverse effects.

Chapter 6 provides further details regarding treatment considerations for the patient's anxiety, including genetic variance and risk of anticholinergic and other GI-related adverse effects.

Summary of Recommendations

- Change aspirin to enteric-coated formulation.
- Treat anxiety according to recommendations made in Chap. 6.
- Assess for and address other possible causes for nausea. If medication is needed, start ondansetron at the lowest effective dose and carefully monitor and titrate.
- For IBS-D, a trial of loperamide or rifaximin may have the best risk-benefit ratio. If this is ineffective or problematic adverse effects such as nausea prohibit the completion of the recommended two-week regimen, alternative therapies such as eluxadodine or bile acid sequestrants (e.g., cholestyramine), should be tried before a TCA such as amitriptyline due to the risk of adverse effects.

- If all other options are exhausted and the patient's goals are still not met, a TCA could be considered at the lowest effective dose after discussing with the patient, considering the higher risks in older adults, but good evidence of benefit with lower doses.

What Did PGx Change About This Case

- Following the clinical decision to use ondansetron instead of metoclopramide, CPIC guidelines reinforced the appropriateness of ondansetron PGx dosing despite the role of CYP2D6 in its metabolism and the presence of an altered CYP2D6 phenotype.
- Proper clinical decision-making identified TCAs as a potential therapeutic option for IBS-D, but CPIC guidelines provided PGx dosing guidance that was indication- and dose-specific and clarified that PGx-focused dose adjustments are not necessary in the context of a low-dose approach to IBS-D management.

References

1. 2023 American Geriatrics Society Beers Criteria® Update Expert Panel. American Geriatrics Society 2023 updated AGS Beers Criteria® for potentially inappropriate medication use in older adults. J Am Geriatr Soc. 2023;71(7):2052–81. https://doi.org/10.1111/jgs.18372. Epub 2023 May 4.
2. An R, Wilms E, Masclee AAM, Smidt H, Zoetendal EG, Jonkers D. Age-dependent changes in GI physiology and microbiota: time to reconsider? Gut. 2018;67(12):2213–22. https://doi.org/10.1136/gutjnl-2017-315542. Epub 2018 Sep 7.
3. Bell GC, Caudle KE, Whirl-Carrillo M, et al. Clinical Pharmacogenetics Implementation Consortium (CPIC) guideline for CYP2D6 genotype and use of ondansetron and tropisetron. Clin Pharmacol Ther. 2017;102(2):213–8. https://doi.org/10.1002/cpt.598. Epub 2017 Apr 6. PMID: 28002639; PMCID: PMC5479760.
4. Dumic I, Nordin T, Jecmenica M, Lalosevic MS, Milosavljevic T, Milovanovic T. Gastrointestinal tract disorders in older age. Can J Gastroenterol Hepatol. 2019;2019:6757524. https://doi.org/10.1155/2019/6757524. eCollection 2019.
5. Hicks JK, Sangkuhl K, Swen JJ, et al. Clinical Pharmacogenetics Implementation Consortium guideline (CPIC) for CYP2D6 and CYP2C19 genotypes and dosing of tricyclic antidepressants: 2016 update. Clin Pharmacol Ther. 2017;102(1):37–44. https://doi.org/10.1002/cpt.597. Epub 2017 Feb 13. PMID: 27997040; PMCID: PMC5478479.
6. Hutchison LC, Sleeper RB, Asal NJ, editors. Fundamentals of geriatric pharmacotherapy: an evidence-based approach. 3rd ed. ASHP Publications; 2025.
7. Kekki M, Samloff IM, Ihamäki T, Varis K, Siurala M. Age- and sex-related behaviour of gastric acid secretion at the population level. Scand J Gastroenterol. 1982;17(6):737–43. https://doi.org/10.3109/00365528209181087.
8. Lau SWJ, et al. Application of pharmacokinetics to specific populations: geriatric, obese, and pediatric patients. In: Shargel L, Yu ABC, editors. Applied biopharmaceutics & pharmacokinetics. 7th ed. McGraw Hill; 2016.

9. Lima JJ, Thomas CD, Barbarino J, et al. Clinical Pharmacogenetics Implementation Consortium (CPIC) guideline for CYP2C19 and proton pump inhibitor dosing. Clin Pharmacol Ther. 2021;109(6):1417–23. https://doi.org/10.1002/cpt.2015. Epub 2020 Sep 20. PMID: 32770672; PMCID: PMC7868475.
10. Lombard DB, Miller RA, Pletcher SD. Biology of aging and longevity. In: Halter JB, Ouslander JG, Studenski S, High KP, Asthana S, Supiano MA, Ritchie C, editors. Hazzard's geriatric medicine and gerontology. 7th ed. McGraw-Hill Education; 2017.
11. Lembo A, Sultan S, Chang L, Heidelbaugh JJ, Smalley W, Verne GN. AGA clinical practice guideline on the pharmacological management of irritable bowel syndrome with diarrhea. Gastroenterology. 2022;163(1):137–51. https://doi.org/10.1053/j.gastro.2022.04.017.
12. Metoclopramide [package insert]. Parsippany: Teva Pharmaceuticals; 2021.
13. Strand DS, Kim D, Peura DA. 25 Years of proton pump inhibitors: a comprehensive review. Gut Liver. 2017;11(1):27–37. https://doi.org/10.5009/gnl15502. PMID: 27840364; PMCID: PMC5221858.
14. Xifaxan [package insert]. Raleigh: Salix Pharmaceuticals, Inc.; 2011.

Chapter 5
Pharmacogenomics and Cardiovascular Care for Older Adults

Casey Rowe, Elina Delgado, Yee Ming Lee, David Bright, and Youssef Roman

Introduction

The prevalence of cardiovascular disease (CVD) among older adults continues to rise as the population ages [2, 43]. Specifically, heart disease has been the leading cause of death for over 100 years, a trend that can likely be attributed to the profound impact of aging on the cardiovascular system [27]. Older adults, specifically those aged 75 years or more, constitute about 30–40% of those hospitalized for ACS and have disproportionally higher mortality rates [7]. The cardiovascular system, like other organ systems, is deeply impacted by the cumulative effects of aging that leads to significant structural and functional changes [13]. The presence of geriatric syndrome is correlated with elevated cardiovascular risk scores, highlighting the interplay between healthy aging and cardiovascular disease burden [4].

C. Rowe (✉)
College of Pharmacy, University of Florida, Gainesville, FL, USA
e-mail: casey.rowe@cop.ufl.edu

E. Delgado
School of Pharmacy, William Carey University, Hattiesburg, MS, USA

Y. M. Lee
Associate Professor of Clinical Pharmacy, Skaggs School of Pharmacy and Pharmaceutical Sciences, University of Colorado, Aurora, CO, USA

D. Bright
Professor of Pharmaceutical Sciences, College of Pharmacy, Ferris State University, Big Rapids, MI, USA

Y. Roman
L.S. Skaggs College of Pharmacy, Idaho State University, Pocatello, ID, USA

T. DeLellis et al. (eds.), *Applying Pharmacogenomics to Optimize Geriatric Care*, https://doi.org/10.1007/978-3-032-10653-7_5

As the cardiovascular system undergoes remodeling with age, several key physiological changes occur. One of the most notable changes is the loss of arterial elasticity due to arterial thickening, increased collagen deposits, and elastic fibers, which leads to increased arterial stiffness [19, 22]. This stiffness contributes to higher systolic blood pressure and decreased coronary perfusion pressure, thereby placing a greater workload on the heart; however, diastolic blood pressure inversely decreases with severity of arterial stiffness. Additionally, the myocardium undergoes hypertrophy and fibrosis, reducing the heart's ability to fill and pump efficiently, thereby increasing the risk for heart failure and atrial fibrillation [32]. Cardiac output and myocardial perfusion are also negatively impacted by a decreased response to beta-adrenergic stimulation in the myocardium and in the periphery (baroreceptors), contributing to impaired blood pressure regulation and increasing the risk of orthostatic hypotension and falls [42] (Atkins 2018). The incidence of valvular disease also increases with age due to age-related changes to fibrosis and progression to stenosis. Aging also impairs endothelial function and nitric oxide synthase activity, affecting vascular tone and increasing the risk of vascular disease and non-ST-elevation MI (NSTEMI) [10].

Cardiovascular care for older adults is often marked by the presence of multiple chronic conditions, which can significantly complicate disease management. Common comorbidities include diabetes, dyslipidemia, and chronic kidney disease. When managing these conditions, pharmacists must carefully consider the physiological changes that occur with aging, as well as the potential for adverse, drug–drug, and drug–disease interactions [10]. For instance, when treating hypertension in patients with heart failure, it is essential to carefully choose antihypertensive agents to avoid exacerbating heart failure symptoms. Similarly, managing diabetes in patients with CVD requires balancing glycemic control with cardiovascular risk reduction. This may involve using medications like SGLT2 inhibitors or GLP-1 receptor agonists, which have been shown to have cardiovascular benefits [33]. Additionally, renal impairment, a common issue in older adults, can also affect the pharmacokinetics of many cardiovascular drugs, necessitating dose adjustments and careful monitoring [32]. The AGS Beers Criteria® highlights several cardiovascular medications that may be potentially inappropriate for the elderly, exacerbate existing conditions, or should be used with extra caution [1].

Pharmacogenomics offers a promising approach for pharmacists to personalize cardiovascular treatment in older adults as it provides another piece of information to guide clinical decision-making. Understanding how patients' genetic variations affect drug metabolism and response allows providers to tailor therapies to optimize drug efficacy and minimize side effects for patients. The Clinical Pharmacogenetics Implementation Consortium (CPIC) has published pharmacogenetic guidelines for clopidogrel, statins, beta-blockers, and warfarin. This chapter will delve into some of the key pharmacogenomics concepts for these drugs followed by case studies that will illustrate the clinical application of pharmacogenomics in optimizing cardiovascular care for the aging population.

Pharmacogenomics of Clopidogrel: The Role of *CYP2C19*

Clopidogrel is a prodrug that requires hepatic biotransformation to form an active metabolite that irreversibly binds to P2Y12 platelet receptors to inhibit platelet aggregation [21]. Only 15% of the prodrug is converted to this active metabolite, while the remaining 85% is hydrolyzed by carboxylesterase-1 (CES1) to inactive forms. Conversion of clopidogrel to its active metabolite involves two sequential oxidative steps involving several CYP450 enzymes (e.g., CYP1A2, CYP2B6, CYP2C9, CYP2C19, and CYP3A4/5), with CYP2C19 having the greatest contribution of all of these enzymes in both steps. The *CYP2C19* genotype has been shown to affect clopidogrel active metabolite formation. Several prospective randomized clinical trials, multicenter pragmatic studies, and meta-analyses have demonstrated the clinical utility of *CYP2C19*-guided antiplatelet therapy [21]. The American Heart Association issued a scientific statement in 2024 supporting *CYP2C19* genetic testing prior to oral P2Y12 inhibitor prescribing in patients with acute coronary syndromes (ACS) or percutaneous coronary intervention (PCI) [29]. Specifically, patients who are CYP2C19 intermediate or poor metabolizers receiving clopidogrel have reduced platelet inhibition and increased risk for developing major adverse cardiovascular and cerebrovascular events.

Based on CYP2C19 phenotype, the CPIC guideline recommends different approaches for clopidogrel use in cardiovascular versus neurovascular settings (detailed in Tables 5.1 and 5.2, respectively). When treating cardiovascular conditions (e.g., ACS/PCI), alternative antiplatelet therapy (like prasugrel or ticagrelor, barring contraindications) is strongly recommended over clopidogrel for patients identified as *CYP2C19* intermediate (IM) or poor metabolizers (PM). For neurovascular indications (e.g., stroke management), CPIC recommends avoiding clopidogrel in poor metabolizers and using an alternative P2Y12 inhibitor; for intermediate metabolizers, considering alternative therapy is moderately recommended. The FDA package insert also includes a boxed warning that notes clinicians should consider an alternate P2Y12 inhibitor in patients with a *CYP2C19* poor metabolizer phenotype [24].

The FDA label also recommends avoiding the concomitant use of clopidogrel with proton pump inhibitors (PPI) such as omeprazole and esomeprazole, which

Table 5.1 CPIC recommendations for Clopidogrel based on CYP2C19 phenotype for cardiovascular indications [21]

CYP2C19 phenotype	CPIC recommendations
CYP2C19 ultrarapid metabolizer	Use standard dose of clopidogrel
CYP2C19 rapid metabolizer	
CYP2C19 normal metabolizer	
CYP2C19 intermediate metabolizer	Avoid clopidogrel if possible. Use prasugrel or ticagrelor if no contraindication
CYP2C19 poor metabolizer	

Table 5.2 CPIC recommendations for Clopidogrel based on CYP2C19 phenotype for neurovascular indications[a] [21]

CYP2C19 phenotype	CPIC recommendations
CYP2C19 ultrarapid metabolizer	No recommendation due to limited data
CYP2C19 rapid metabolizer	
CYP2C19 normal metabolizer	Use standard dose of clopidogrel
CYP2C19 intermediate metabolizer	Consider an alternative P2Y12 inhibitor if clinically indicated and no contraindication.
CYP2C19 poor metabolizer	Avoid clopidogrel if possible. Use ticagrelor or ticlopidine. Avoid prasugrel in patients with a history of stroke or transient ischemic attack.

[a]Includes acute ischemic stroke or transient ischemic attack, secondary prevention of stroke, or prevention of thromboembolic events following neurointerventional procedures, such as carotid artery stenting and stent-assisted coiling of intracranial aneurysms

could reduce CYP2C19 activity. Other PPI such as lansoprazole and pantoprazole are less potent CYP2C19 inhibitors and hence have less impact on the antiplatelet activity of clopidogrel. This drug–drug–gene interaction results in CYP2C19 phenoconversion, where there is a mismatch between genotype-predicted phenotype and actual phenotype. While CYP2C19 phenoconversion can occur, it remains unclear if this clopidogrel–PPI interaction has any real-world clinical implications [38]. Further studies are needed to determine the clinical outcome of this interaction.

Pharmacogenomics of Statins: The Role of *SLCO1B1*, *ABCG2*, and *CYP2C9*

Several key genes influence statin pharmacokinetics and the risk of adverse effects. The *SLCO1B1* gene encodes the hepatic uptake transporter OATP1B1 (also called SLCO1B1), which plays a critical role in facilitating the transport of all statins into the liver. Genetic variants in *SLCO1B1* that result in decreased or poor OATP1B1 transporter function leading to increased systemic statin exposure and the risk of statin-associated muscle symptoms (SAMS), a known dose-dependent toxicity. Recognizing this, the CPIC guideline provides specific statin and dose recommendations with the respective SAMS risk based on SLCO1B1 transporter function in algorithmic form [6]. The FDA labels of atorvastatin, rosuvastatin, and simvastatin mention the role of OATP1B1 in the disposition of these statins, but provide no pharmacogenomic dosing recommendations [5, 34, 36].

Table 5.3 CPIC dosing recommendations for rosuvastatin based on ABCG2 phenotype [6]

Phenotype	CPIC recommendations
ABCG2 normal function	Prescribe desired rosuvastatin starting dose
ABCG2 decreased function	
ABCG2 poor function	Prescribe rosuvastatin ≤20 mg as starting dose. If dose >20 mg needed, consider an alternative statin or combine with a non-statin medical therapy.

For certain statins, additional genetic factors are more relevant. Rosuvastatin disposition, for instance, is significantly modulated by the ABCG2 transporter (also known as BCRP), encoded by the *ABCG2* gene. CPIC guidance includes specific rosuvastatin dose recommendations based on SLCO1B1 phenotype alone, ABCG2 phenotype alone (Table 5.3) and combinations of *SLCO1B1* and *ABCG2* phenotypes (see Table 5.4) [6]. Of note, the FDA label of rosuvastatin has a race-based recommendation where a lower rosuvastatin starting dose of 5 mg is recommended for patients of Asian biogeographical ancestry groups [34]. This is because pharmacokinetic studies demonstrated an almost twofold higher exposure among Asian subjects than Caucasian controls. A recent study that evaluated the frequency ABCG2 and SLCO1B1 variants in Asian, Native Hawaiian, and Pacific Islander women found a lower frequency of SLCO1B1 variant but a higher frequency of ABCG2 variant in this cohort (13–46%) compared to Europeans (9.4%) [3]. These findings may account for FDA's empiric recommendation to use a lower rosuvastatin starting dose in patients of Asian biogeographical ancestry groups. The rosuvastatin starting dose can now be further tailored based on patient's specific ABCG2 and SLCO1B1 phenotypes instead of being race-based. While providers consider pharmacogenomic results in dosing rosuvastatin, they need to consider nongenetic variables such as renal impairment in older adults, which would limit the starting rosuvastatin dose due to its renal clearance. For fluvastatin, its metabolism is primarily mediated by the CYP2C9 enzyme. Genetic variants leading to reduced CYP2C9 activity are associated with increased fluvastatin exposure and a consequently higher risk of SAMS. Therefore, CPIC also includes fluvastatin dose recommendations for the combination of *CYP2C9* and *SLCO1B1* phenotypes (see Table 5.5).

Table 5.4 CPIC dosing recommendations for rosuvastatin based on ABCG2 and SLCO1B1 phenotypes [6]

	ABCG2 normal function	ABCG2 decreased function	ABCG2 poor function
SLCO1B1 increased function	Prescribe desired rosuvastatin starting dose		Prescribe rosuvastatin ≤20 mg per day as starting dose. If dose >20 mg is needed, consider an alternative statin or combine with a non-statin.
SLCO1B1 normal function			
SLCO1B1 decreased function	Prescribe desired rosuvastatin starting dose. Be aware of possible increased risk for myopathy for doses >20 mg per day.		Prescribe rosuvastatin ≤10 mg per day as starting dose. If dose >10 mg is needed, consider an alternative statin or combine with a non-statin medical therapy.
SLCO1B1 poor function	Prescribe rosuvastatin ≤20 mg per day as starting dose. If dose >20 mg needed, consider alternative statin or combine with a non-statin medical therapy.		Prescribe rosuvastatin ≤10 mg per day as starting dose. If dose >10 mg is needed, combine with a non-statin medical therapy.

Table 5.5 CPIC recommendations for fluvastatin based on CYP2C9 and SLCO1B1 phenotypes [6]

	CYP2C9 normal metabolizer	CYP2C9 intermediate metabolizer	CYP2C9 poor metabolizer
SLCO1B1 increased function	Prescribe desired starting dose.	Prescribe ≤40 mg per day as starting dose. If dose >40 mg is needed, consider an alternative statin or combine with a non-statin medical therapy.	Prescribe ≤20 mg per day as starting dose. If dose >20 mg is needed, consider an alternative statin or combine with a non-statin medical therapy.
SLCO1B1 normal function			
SLCO1B1 decreased function	Prescribe desired starting dose. Be aware of possible increased risk for myopathy for doses >40 mg per day.	Prescribe ≤20 mg per day as starting dose. If dose >20 mg is needed, consider an alternative statin or combine with a non-statin medical therapy.	Use an alternative statin.
SLCO1B1 poor function	Prescribe ≤40 mg per day as starting dose. If dose >40 mg is needed, consider an alternative statin or combine with a non-statin medical therapy.	Use an alternative statin.	Use an alternative statin.

Table 5.6 CPIC recommendations for metoprolol based on CYP2D6 phenotype [12]

Phenotype	Activity score	CPIC recommendations
CYP2D6 ultrarapid metabolizer	>2.25	No recommendation due to limited data
CYP2D6 normal metabolizer	$1.25 \leq X \leq 2.25$	Use standard metoprolol dose
CYP2D6 intermediate metabolizer	$0 < X < 1.25$	
CYP2D6 poor metabolizer	0	Initiate with the lowest metoprolol starting dose and carefully titrate the dose upward. Monitor closely for bradycardia. Alternatively, consider using another beta-blocker.

Pharmacogenomics of Beta-Blockers: The Role of *CYP2D6*

Certain beta-blockers are metabolized via the CYP2D6 enzyme with data demonstrating a pharmacokinetic association between *CYP2D6* phenotype and beta-blocker exposure. To-date, the strongest evidence has been reported for *CYP2D6* poor metabolizers and the impact on metoprolol plasma levels and heart rate [12]. Of note, most of the data reported for the association between *CYP2D6* genotype and metoprolol response relate to oral metoprolol formulations with limited data for intravenous formulations. CPIC guidelines recommend initiating metoprolol at the lowest starting dose in patients who are *CYP2D6* poor metabolizers and carefully titrating the dose with close monitoring for bradycardia (see Table 5.6). However, recognizing that many patients achieve steady state within days of starting metoprolol, it should be emphasized that dose adjustment recommendations pertain only to starting doses, and that those who have long been on a stable dose of metoprolol likely need no further adjustment if clinical markers (e.g., blood pressure, heart rate) meet their appropriate thresholds [12]. The FDA label for metoprolol reports that CYP2D6 poor metabolizers have higher blood levels [28]. It also reports a drug interaction when strong CYP2D6 inhibitors (such as quinidine, fluoxetine, and paroxetine) are given to CYP2D6 normal metabolizers as it results in higher metoprolol levels, decreasing the cardioselectivity of metoprolol.

Pharmacogenomics of Warfarin: The Role of *CYP2C9*, *VKORC1*, *CYP4F2*, and *CYP2C Cluster rs12777823*

Despite the availability of direct oral anticoagulant (DOAC) agents, warfarin is still being utilized in practice, whether that is due to lack of indication for specific patient populations or medication cost. Dosing remains challenging given its narrow therapeutic index and wide interpatient dose variability. Common genetic polymorphisms in *CYP2C9*, *VKORC1*, *CYP4F2*, and *rs12777823* and known nongenetic factors contribute significantly toward warfarin dose variability. Among patients of European biogeographical ancestry groups, common variants in *CYP2C9*, *VKORC1*, and *CYP4F2* account for up to 18%, 30%, and 11% of the variability in warfarin dosing, respectively [16].

CYP2C9 is the primary metabolizing enzyme of the more potent isomer, S-warfarin. The presence of *CYP2C9* *2 or *3 allele reduces warfarin metabolism by 30–40% and 80–90%, respectively. Other *CYP2C9* star alleles (*CYP2C9* *5, *6, *8, and *11) are also associated with reduced CYP2C9 enzyme function. These are more commonly found in individuals of African biogeographical ancestry groups and are sometimes called African-specific *CYP2C9* alleles. As most labs test for *CYP2C9* *2 and *3, one must check if the lab also tests for African-specific alleles of *CYP2C9* *5, *6, *8, and *11 in a patient who self-reports to be of an African biogeographical ancestry group. Otherwise, one may overestimate the warfarin dose for these patients. As *VKORC1* gene encodes for warfarin's target, *VKORC1* genetic polymorphism is associated with altered warfarin sensitivity. CYP4F2 catalyzes the metabolism of vitamin K from the vitamin K cycle. Genetic polymorphism of *CYP4F2* modestly impacts warfarin dosing. A *CYP2C* cluster rs12777823 SNP was identified in a study among African Americans that was associated with warfarin dose reduction among patients of African biogeographical ancestry groups by about 7–9 mg per week. Of note, current data do not suggest an association between rs12777823 variant and warfarin dose in non-African Americans.

To account for African-specific alleles based on self-identified ancestry, CPIC guidelines contain an algorithm that stratifies the warfarin dosing process by taking into consideration the patient's self-identified ancestry [16]. There are warfarin dosing calculators online such as warfarindosing.org that allow for the input of patient's genetic (e.g., CYP2C9, VKORC1) and nongenetic factors (e.g., age, gender, weight, concomitant drugs) into an algorithm to predict an initial warfarin dose.

Case 1

An ambulatory 73-year-old male with a history of hypertension and heart failure with reduced ejection fraction (HFrEF) (ACC/AHA Stage C, NYHA Class II) presents to the outpatient cardiology clinic following a STEMI 1 week ago. The STEMI was managed with a percutaneous coronary intervention (PCI) and the placement of a drug-eluting stent. This visit is a follow-up to assess if any further medication adjustments are appropriate following the STEMI. The patient's pharmacogenomic results were previously ordered and uploaded into the electronic medical record.

Medical History:

Hypertension (HTN; 22 years)
Dyslipidemia (22 years)
Type 2 Diabetes (T2D; 14 years)
HFrEF (ACC/AHA Stage C, NYHA Class II; 3 years)
Chronic Kidney Disease (CKD) Stage 3a (5 years)
Benign Prostatic Hyperplasia (BPH; 3 years)
STEMI (DES × 2; 1 week ago)
Seasonal allergies (70 years)

Allergies:

Penicillin (hives)

Social History:

Lives in a house with wife of 50 years; drinks a glass of wine with dinner a few times a week; former smoker

Pertinent vitals and labs:

Today: Height 5′11″, weight 212 lbs, BMI 29.6 kg/m^2, BP 118/74 mmHg, HR 64 BPM, eGFR 52 mL/min, EF = 37%
Fasting lipid panel (2 days ago): TG 120 mg/dL, HDL 54 mg/dL, LDL 72 mg/dL, TC 150 mg/dL
A1C 8.2%

Pharmacogenomic Results:

*CYP2C19 *2/*3* Poor metabolizer
*CYP2D6 *1/*5* Intermediate metabolizer
*CYP2C9 *1/*1* Normal metabolizer
SLCO1B1 Normal function
ABCG2 Normal function

Current medications:

Aspirin 81 mg po daily (new)
Clopidogrel 75 mg po daily (new)
Furosemide 20 mg po daily
Glipizide ER 10 mg po daily
Lisinopril 10 mg po daily
Loratadine/pseudoephedrine 10 mg/240 mg (Claritin-D 24-h) po daily
Metformin 1000 mg po twice daily
Metoprolol succinate XL 25 mg po daily
Rosuvastatin 20 mg po daily
Nitroglycerin 0.4 mg SL PRN
Spironolactone 25 mg po daily
Tamsulosin 0.4 mg po daily

Clinical (Nongenetic) Considerations

The patient's treatment goals include optimizing secondary ASCVD prevention following the STEMI, refining heart failure therapy, and managing cardiovascular risk factors. Addressing the care of a 73-year-old patient with complex comorbidities including hypertension, HFrEF, and a STEMI sustained 1 week ago requires careful evaluation of specific geriatric factors. Aging is associated with a growing burden of cardiovascular disease, and this patient's complex medical history necessitates careful consideration of his multiple comorbidities, including CKD, diabetes, and BPH. In addition to these comorbidities, polypharmacy is a concern, as the patient is currently taking multiple medications. When evaluating treatment options, it is essential to consider the potential for drug–drug interactions, drug–gene interactions, adverse drug reactions, and the impact of age-related changes on drug pharmacokinetics and pharmacodynamics.

This patient's previous cardiovascular history and risk factors, coupled with the recent STEMI, place him at increased risk for future cardiovascular events, especially within the first year following the STEMI. Guideline-recommended pharmacotherapy following STEMI and PCI include:

- Antithrombotic therapy
- Beta-blockers
- Renin-angiotensin-aldosterone system (RAAS) antagonists
- Statins
- As needed, a fast-acting nitrate

Long-term management after acute coronary syndrome (ACS) is focused on decreasing atherogenic risk and myocardial stress. Dual antiplatelet therapy (DAPT) with aspirin and a P2Y12 inhibitor (clopidogrel, prasugrel, or ticagrelor) for 1 year is currently considered the standard of care following a STEMI. For patients who are deemed to have a high risk for bleeding, monotherapy with a P2Y12 inhibitor may be considered after 1–3 months. Within 24 hours of his STEMI, the patient was initiated on clopidogrel and aspirin to provide antiplatelet therapy. High-intensity statin initiation is also currently recommended to achieve an LDL reduction greater than 50% from baseline with a goal LDL less than 55 mg/dL. The patient was established on rosuvastatin 20 mg prior to his hospitalization. His most recent LDL was 72 mg/dL, higher than the goal. Considerations on treatment to achieve LDL goals include statin tolerance (dose-related SAMS risk), renal function, metabolism, and addition of a medication from a different class. For this patient, options would include increasing rosuvastatin to 40 mg, adding ezetimibe, or a PCSK9 inhibitor. These pharmacotherapeutic agents should always be coupled with lifestyle modifications (i.e., regular physical activity, heart-healthy diet, limited alcohol consumption), as appropriate for the patient.

Beta-blocker therapy after ACS should be initiated within 24 hours, unless a contraindication exists. For this patient, beta blocker selection is going to be guided by his concomitant HFrEF and hypertension. Guideline-recommended medical therapy for Stage C HFrEF consists of an Angiotensin Receptor-Neprilysin Inhibitor (ARNI), evidence-based beta-blocker (metoprolol XL, bisoprolol, or carvedilol), mineralocorticoid antagonist, an SGLT2-inhibitor, and a diuretic as needed for volume overload. Although the target dose for metoprolol succinate in HFrEF is 200 mg daily, in the elderly, this needs to be balanced with the risk for bradycardia, masking of hypoglycemia, and increased risk for orthostatic hypotension. Today, the patient's blood pressure is within normal limits, and his heart rate is at the goal of less than 70 BPM.

Preferential use of an ARNI over an angiotensin-converting enzyme inhibitor (ACEi) or an angiotensin-receptor blocker (ARB) should be considered for this patient unless he reports an intolerance, contraindication, or inaccessibility to sacubitril/valsartan. Currently, the patient's RAAS inhibition is managed with the ACE inhibitor, lisinopril. When transitioning a patient to sacubitril/valsartan from an ACEi, there is a required 36-hour washout period, and the recommended initial dose for this patient would be sacubitril 24 mg/valsartan 26 mg twice daily. Blood pressure, renal function, and potassium should be monitored closely following initiation and titration to the target dose. Consideration should also be given to the addition of an SGLT-2 inhibitor (dapagliflozin, empagliflozin or sotagliflozin) given the patient's HFrEF, CKD, and diabetes status. Since SGLT-2 inhibitors have diuretic effects by eliminating glucose in the urine, the patient should be counseled on monitoring volume status to prevent hypovolemia/hypotension and the increased risk of urogenital infections. Further evaluation of the patient's medication regimen for HFrEF shows that he is on all other recommended therapies (furosemide and spironolactone).

Given the patient's age and ambulatory status, the systolic blood pressure goal for this patient is <130 mmHg [18, 41]. At clinic today, his BP is within goal at 118/74 mmHg. The patient's concomitant disease states will guide medication selection and management of his hypertension. For patients who do not have good blood pressure control and require a beta-blocker, carvedilol or labetalol can be considered. Given his age and the use of multiple medications that have antihypertensive effects, the patient is at increased risk for orthostatic hypotension. Careful titration and monitoring are required.

Glycemic control is imperative to reduce cardiovascular risk in patients with diabetes. The patient's most recent A1c is 8.2% on metformin and glipizide, above the goal of less than 7% set by the ADA and cardiovascular guidelines. Considerations in the selection of antidiabetic agents for this patient include the presence of CVD, CKD, and age. Provided there are no limitations to the patient initiating an SGLT-2, the addition of this agent will assist with A1c reduction on average by 0.5–1% [44]. Glipizide, a sulfonylurea, should be used with caution in older adults due to increased risk of hypoglycemic events and lack of evidence to support cardiovascular or renal protective effects [1]. The goal would be to shift away from glipizide use for this patient, and the speed at which that is done is variable based on signs and symptoms, insurance coverage of other agents, and patient preference. In the event the patient is unable to meet the A1c goal with an SGLT-2 inhibitor, the next recommendation would be the addition of a GLP-1 agonist to assist with cardiovascular and renal outcomes.

The patient's tamsulosin, a selective alpha-1a blocker, has been used to manage his benign prostatic hyperplasia (BPH) symptoms. This medication has not had a significant impact on his hypertension treatment plan, but its potential to cause orthostatic hypotension should still be considered. To minimize the risk of orthostatic hypotension, the patient's blood pressure should be carefully monitored, and adjustments to his hypertension treatment plan may be necessary.

Additionally, when considering that the patient is also taking loratadine/pseudoephedrine to treat his allergies, it is essential to assess the potential drug-disease interactions. The pseudoephedrine component can increase the risk of urinary retention, which could exacerbate the patient's BPH symptoms. Older adults may have an increased sensitivity to the sympathomimetic properties of pseudoephedrine given concomitant disease states (CVD and diabetes) and polypharmacy. Given the risk of worsening his BPH and potential to increase the risk of cardiovascular events, it may be beneficial to consider alternative allergy medications that do not contain pseudoephedrine. Loratadine, cetirizine and fexofenadine are antihistamines that can relieve allergy symptoms without the pseudoephedrine component. It would also be reasonable to consider topical therapies (intranasal corticosteroids or antihistamines) to avoid the risk of systemic adverse effects.

Pharmacogenomic Considerations

The patient's pharmacogenomic results indicate poor CYP2C19 metabolism phenotype, which is associated with reduced clopidogrel active metabolite formation and decreased antiplatelet response. According to the CPIC guidelines and FDA boxed warning, clopidogrel is not recommended for this patient due to the increased risk of major adverse cardiovascular and cerebrovascular events, and an alternative antiplatelet should be used [21, 24]. Alternatives include prasugrel (considering age, weight, and lack of previous stroke/TIA) or ticagrelor. The patient should be transitioned to prasugrel 10 mg PO daily or ticagrelor 90 mg PO twice daily with awareness that these agents have been shown to have increased bleeding rates in older adults [1]. Should the patient require treatment at the age of 75 years or older, prasugrel should be reevaluated and dose either reduced to 5 mg PO daily or discontinued. Aspirin 81 mg is an appropriate dose and should be continued.

The patient exhibits a CYP2D6 intermediate metabolizer phenotype and may have a slightly higher metoprolol exposure that may impact his heart rate and achievement of target HF dosing. CPIC guidelines recommend slow dose initiation of metoprolol in CYP2D6 poor metabolizer phenotypess with close monitoring for bradycardia, but to initiate standard dose in those expressing a CYP2D6 intermediate metabolizer phenotype [12]. Since the patient is stable on the current dose (5 or more days) with HR at goal, no dose adjustment is necessary.

The patient has normal SLCO1B1 and ABCG2 function, suggesting a normal risk for SAMS. Given his recent ACS event and concomitant high-risk comorbidities (diabetes, CKD) necessitating high-intensity treatment, increasing rosuvastatin to 40 mg daily is clinically appropriate. Considering the patient's pharmacogenomic results and adequate renal function (eGFR), increasing rosuvastatin to the maximum dose is not contraindicated at this time.

Summary of Recommendations

- Transition from clopidogrel to prasugrel 10 mg PO daily or ticagrelor 90 mg PO twice daily, considering the potential for increased bleeding risk.
- Continue aspirin, metoprolol succinate, furosemide, metformin, and spironolactone.
- Consider the addition of an SGLT-2 inhibitor (dapagliflozin, empagliflozin, or sotagliflozin) for HFrEF, CKD, and diabetes management and discontinuation of glipizide.
- Consider preferential use of an ARNI (sacubitril/valsartan) over an ACEi or ARB, unless contraindicated or inaccessible.
- Consider increasing rosuvastatin to the higher dose of 40 mg po daily, despite 20 mg being considered high intensity as well.
- Consider alternative treatments for allergic rhinitis to deprescribe pseudoephedrine.
- Monitor home blood pressure and orthostasis symptoms closely for need to change tamsulosin therapy. Deprescribing pseudoephedrine could help control BPH symptoms if a change in tamsulosin is needed.

What Did Pharmacogenomics Change About the Case Outcome?

- The use of pharmacogenomic testing in this case led to the identification of a poor CYP2C19 metabolizer phenotype, necessitating a change in antiplatelet therapy to reduce the risk of major adverse cardiovascular and cerebrovascular events.
- The integration of PGx testing into the treatment plan allowed for a more personalized approach to medication selection, ultimately improving the patient's treatment outcomes and reducing the risk of adverse events.
- While metoprolol has PGx implications for starting doses, no pharmacogenomic-guided dose adjustments are necessary.
- Pharmacogenomic testing also showed normal SAM risk associated with rosuvastatin use, based on ABCG2 and SLCO1B1 results. Given the clinical need for a change to a high-intensity statin, the pharmacogenomic data provides reassurance that SAM risk is not genetically elevated in this patient.

Case 2

A 75-year-old Caucasian woman with a long history of hypertension and dyslipidemia presents to the clinic with a complaint of myalgia. The patient's physician is concerned that the myalgia may be a statin-related side effect; however, they are hesitant to discontinue the medication given the patient's high risk for cardiovascular events. The physician has ordered pharmacogenomic testing to inform their decision-making process. Your expertise is sought to help interpret the test results and guide management considering the patient's frailty status and potential risks and benefits of statin therapy.

Medical History:

Hypertension (44 years)
Dyslipidemia (37 years)
Depression (5 years)
Osteoporosis (2 years)

Allergies:

None known

Social History:

Current smoker (1 ppd × 40 years); occasional alcohol consumption (socially, 2 drinks per week); retired teacher; lives alone in ground-floor apartment; daughter lives nearby and visits 2–3 times a week to help with groceries and errands

Pertinent Vitals and Labs:

Today: Height 5′3″, weight 112 lb, BP 148/80 mmHg, HR 58 BPM, eGFR 65 mL/min, BMI 19.8 kg/m^2
Lipid panel (3 weeks ago): TG 110 mg/dL, HDL 38 mg/dL, LDL 150 mg/dL, TC 210 mg/dL

Additional Exams:

CAC: 44 (mild)
DEXA scan: Lumbar spine (T-score −1.3), right femoral neck (T-score −2.1), left femoral neck (T-score −2.6) and total hip (T-score −2)
PHQ9 score: 2
Clinical Frailty Scale: 5

Pharmacogenomic Results:

*SLCO1B1 *1/*5* Decreased function
ABCG2 Normal function
*CYP2C19 *1/*1* Normal metabolizer
*CYP2D6 *1/*2* Normal metabolizer
*CYP2C9 *1/*1* Normal metabolizer

Current Medications:

Aspirin 81 mg po daily
Calcium/vitamin D petite supplement po daily
Denosumab 60 mg SubQ every 6 months
Escitalopram 10 mg po daily
Fish oil 1200 mg po daily as a dietary supplement
Hydrochlorothiazide (HCTZ) 25 mg po daily
Simvastatin 40 mg po daily

Clinical (Nongenetic) Considerations

The treatment goals for this patient are to promote healthy aging, maintain functional status, and reduce her risk of future cardiovascular events, considering her mild frailty (CFS 5). The patient's prolonged cardiovascular history, in conjunction with her age and smoking history, places her at an elevated risk of a cardiovascular event.

First, primary prevention of cardiovascular disease can be addressed through LDL reduction. In this case, the patient's estimated 10-year ASCVD risk is 37.2%. This risk level strongly supports the use of a high-intensity statin aimed at achieving at least a 50% reduction in LDL-C from baseline, with a target LDL-C of 100 mg/dL or less. Her current simvastatin 40 mg regimen provides only moderate-intensity lipid lowering (30–49% reduction) and is therefore insufficient to meet this goal. Most importantly, however, simvastatin 40 mg is a likely culprit of the myopathy, particularly when considering her *SLCO1B1* decreased function phenotype that is associated with increased risk for SAMS. Her risk factors for SAMS include older age, female sex, use of high simvastatin dose, and low body mass [40]. Other risk factors such as hypothyroidism and vitamin D deficiency can be verified through laboratory testing and corrected if present.

Hypertension is highly prevalent among older adults and often coexists with frailty, a condition characterized by decreased physiological reserve and increased vulnerability to stressors. In frail older adults like this patient, managing hypertension requires a delicate balance to avoid adverse effects such as orthostatic hypotension and falls. Frailty can complicate hypertension management due to polypharmacy, comorbidities, and reduced adherence to treatment [14, 36]. Blood pressure targets may need to be adjusted to avoid overly aggressive treatment, which can exacerbate frailty and lead to adverse outcomes.

Further medication optimization includes strongly suggesting smoking cessation and offering comprehensive support with counseling and pharmacotherapy, which is a crucial step in reducing her overall cardiovascular risk. The routine use of aspirin 81 mg for primary prevention is generally not recommended for individuals over 70 due to bleeding risks outweighing benefits; therefore, discontinuation should be discussed with the patient. Similarly, the fish oil supplement lacks clear indication and contributes to pill burden and bleeding risk, so it is recommended to discontinue it via a shared discussion with the patient to explain the risk-benefit profile. High-dose omega-3 fatty acid (specifically icosapent ethyl) is indicated to treat hypertriglyceridemia in patients with high ASCVD risk or in patients with Stage C HFrEF, which is not present in this patient. The patient's current treatment for osteoporosis (denosumab, calcium/vitamin D) and depression (escitalopram) are appropriate and should be continued, with ongoing emphasis on fall prevention strategies.

Non-pharmacological recommendations are also important to promote healthy aging in this patient. She should be encouraged to adopt and maintain a heart-healthy diet and engage in regular physical activity appropriate for her level of frailty, focusing on elements like walking, balance, and light strength training to preserve function and reduce fall risk.

Pharmacogenomic Considerations

Pharmacogenomic testing yielded significant results relevant to her statin therapy. Her CYP2C19 and CYP2D6 phenotypes were found to be normal metabolizer phenotypes (**1/*1* and **1/*2*, respectively), indicating that a normal response can be expected to medications metabolized via these pathways such as escitalopram via CYP2C19 (see Chap. 7). The patient has the SLCO1B1 decreased function phenotype, which is strongly associated with increased systemic exposure to simvastatin and a heightened risk of SAMS, providing a probable explanation for the patient's complaint of myalgia.

Given the patient's SLCO1B1 genotype and the high SAMS risk associated with the simvastatin 40 mg dose, it is recommended to discontinue simvastatin and switch to a high-intensity statin at a dose associated with lower risk of SAMS [6, 17]. Based on the CPIC guidelines, which include an algorithm for statin choice based on SLCO1B1 activity, desired intensity, and SAMS risk, her option includes rosuvastatin 20 mg po daily [6]. Considering the patient's mild CAC score and age over 75, a thorough clinical assessment and risk discussion are warranted before making changes to her statin therapy. A washout period such as a week between stopping simvastatin before starting rosuvastatin can be considered to allow her myalgia symptoms to clear [40]. The patient should be monitored closely for any recurrent muscle symptoms. Alternative dosing strategies (e.g., three times a week dosing) could be explored if daily therapy results in any side effects. Alternatively, PCSK9 inhibitors could be considered, although they may not be clinically appropriate at this time. Given her smoking status, smoking cessation should be a priority, and the potential impact of smoking on CYP1A2 metabolism should be considered even though the patient is not currently on a CYP1A2 substrate.

Summary of Recommendations

- The patient's myalgia is likely related to her simvastatin therapy, given her SLCO1B1 decreased function phenotype.
- Switching to a statin with a lower SAMS risk and addressing other aspects of her cardiovascular and overall health will help optimize her lipid treatment and improve her quality of life.

What Did Pharmacogenomics Change About the Case Outcome?

- Pharmacogenomics testing revealed the patient's SLCO1B1 decreased function, which is associated with increased systemic exposure to simvastatin and a heightened risk of SAMS) and provided a probable explanation for the patient's presenting complaint of myalgia as well as reasonable alternate drugs/doses.
- Based on the patient's SLCO1B1 decreased function phenotype, simvastatin should be discontinued, and a high-intensity statin with lower risk of SAMS initiated for primary prevention of ASCVD since her ASCVD 10-year risk is over 20%.

Case 3

A 78-year-old male of Asian biogeographical ancestry group with a significant past medical history of hypertension and Type 2 diabetes (T2D) was admitted to the hospital following an acute ischemic stroke with a NIH stroke scale score of 4 (minor stroke/TIA). The stroke team initiated dual antiplatelet therapy to mitigate the risk of recurrent stroke. You are reviewing patient's medication orders and see that the patient has pharmacogenomic results.

Medical History:

Hypertension (30 years)
Type 2 diabetes (28 years)

Allergies:

None known

Social History:

Lives with wife in assisted living, married for 60 years with two adult children who live nearby; retired accountant; nonsmoker; never used illicit substances; 1 cup of coffee daily

Pertinent Vitals and Labs:

Height 5′5″, weight 150 lb, BMI 25 kg/m2, BP 125/80 mmHg, HR 65 BPM, HbA1c 7.5%, eGFR 60 mL/min
Lipid panel: TG 150 mg/dL, HDL 45 mg/dL, LDL 150 mg/dL, TC 225 mg/dL

Pharmacogenomic Results:

*CYP2C19 *1/*2* Intermediate metabolizer
*CYP2D6 *4/*4* Poor metabolizer
*SLCO1B1 *1/*5* Decreased function
ABCG2 Decreased function

Current Medications:

Aspirin 325 mg po daily (new)
Atorvastatin 80 mg po daily (new)
Clopidogrel 75 mg po daily (new)
Hydrochlorothiazide 25 mg po daily
Lisinopril 10 mg po daily
Metformin 500 mg po BID
Metoprolol succinate XL 25 mg po daily

Clinical (Nongenetic) Considerations

The primary treatment goals for this patient are to prevent recurrent stroke, manage cardiovascular risk factors, and optimize diabetes control [28]. Given the patient's recent stroke, minimization of atherosclerotic risk is imperative and is done through the initiation of DAPT (aspirin and clopidogrel) for 90 days and a high-intensity statin (atorvastatin 40 mg). The patient's blood pressure and diabetes control will also be closely monitored to ensure optimal management.

Following DAPT for 90 days, patients should be continued on lifelong ASA therapy if there is no contraindication. However, the patient is on aspirin 325 mg daily. High aspirin doses (>100 mg) have been reported in the PLATO trial to compromise the effectiveness of ticagrelor [39]. The THALES trial, which investigated the use of ticagrelor and aspirin in acute ischemic stroke patients, employed a maintenance aspirin dose of 75–100 mg po daily [23]. Hence, this further supports the reduction in aspirin dose to 81 mg po daily. Additionally, aspirin at a dose of 325 mg does not provide additional clinical benefit but increases the risk of bleeding [26]. Therefore, reducing the patient's aspirin dose to 81 mg daily is advisable to maintain a favorable balance between efficacy and safety.

At present, there is no one agent or drug class preferentially recommended for hypertension management following an ischemic stroke/TIA. Review of the patient's current medications and past medical history suggests the use of lisinopril, an angiotensin-converting enzyme (ACE) inhibitor, hydrochlorothiazide (HCTZ), a thiazide diuretic, and metoprolol succinate, a beta-blocker, for management of hypertension. Given the lack of morbidity and mortality benefits with the use of beta blockers for hypertension management, it is recommended that the beta-blocker be discontinued. It is recommended that the patient continue lisinopril and HCTZ to maintain optimal blood pressure control, as uncontrolled hypertension is a significant risk factor for recurrent stroke. Emphasize the importance of strict blood pressure control, with a target blood pressure of less than 130/80 mmHg, to minimize the risk of future cerebrovascular events. Should the blood pressure increase after removal of low-dose metoprolol succinate, consideration could be given to increasing the dose of lisinopril or transitioning to chlorthalidone from HCTZ.

The patient's elevated A1c level of 7.5% indicates that their diabetes is not under control. Metformin should be increased to 1000 mg BID as the patient's estimated glomerular filtration rate (eGFR) is greater than 30 mL/min but a secondary agent should be considered to optimize glycemic control. In this case, it would be appropriate to initiate a SGLT2 inhibitor, as it is the preferred option in the absence of a compelling indication for a GLP-1 agonist, such as obesity (BMI 25 kg/m^2). The addition of an SGLT2 inhibitor will help to improve glycemic control, reduce cardiovascular risk, and mitigate the progression of kidney disease [22].

Pharmacogenomics Considerations

The patient has a CYP2C19 intermediate metabolizer phenotype; therefore, clopidogrel is not recommended for this patient due to the reduced efficacy associated with CYP2C19 intermediate metabolizer phenotypes, which reduces the conversion of clopidogrel to its active metabolite. According to the clopidogrel CPIC guidelines for neurovascular indications, there is an increased risk for major adverse cardiovascular and cerebrovascular events when treated with clopidogrel; therefore, it is recommended to use an alternative antiplatelet instead to mitigate this risk. However, prasugrel is contraindicated in patients with a history of stroke due to the increased risk of bleeding associated with this medication [21]. Ticagrelor is an optimal option because it is not affected by the CYP2C19 pathway. Therefore, it is recommended that the patient switch from clopidogrel to ticagrelor 90 mg po twice daily.

This patient has newly started on high-intensity atorvastatin 80 mg daily for secondary ASCVD prevention. However, the patient has SLCO1B1 decreased function which is associated with high SAMS risk when prescribed high-intesnity atorvastatin. An alternative high-intensity statin with low SAM risk based on CPIC guideline for patients with SLCO1B1 decreased function phenotype is rosuvastatin 20 mg daily [6]. Although the patient's eGFR does not preclude the use of rosuvastatin 20 mg dose, the reduced ABCG2 function, consistent with increased risk in patients of Asian biogeographic ancestry, makes a case for pre-emptively changing to rosuvastatin to preserve statin use. Where the patient is already taking and tolerating the atorvastatin 80 mg daily dose, initiating a therapy change based on phenotype alone in the absence of anticipated drug–drug or drug–disease interactions may not be warranted. However, since the phenotype was obtained before initiating atorvastatin therapy, it would be reasonable to initiate rosuvastatin instead to preserve high-intensity statin therapy given his strong indication. Waiting to see if atorvastatin is tolerated, in a patient with these risk factors could lead to severe adverse effects resulting in avoidance of statins altogether.

Summary of Recommendations

- Switch from clopidogrel to ticagrelor 90 mg po twice daily.
- Reduce aspirin dose to 81 mg po daily.
- Consider an alternative statin (rosuvastatin 20 mg po daily) due to increased risk of SAMS on atorvastatin 80 mg po daily dose.
- To discontinue metoprolol given there is no compelling indication and consider alternative blood pressure medications as clinically appropriate.

What Did Pharmacogenomics Change About the Case Outcome?

- The patient's *CYP2C19 *1/*2* intermediate metabolizer diplotype necessitated a switch from clopidogrel to ticagrelor (per CPIC guidelines) to ensure optimal antiplatelet therapy. This also led to decreasing the aspirin dose to avoid its interaction with ticagrelor and reduce the bleeding risk.
- The patient's *SLCO1B1 *1/*5* decreased function diplotype and *ABCG2* decreased function increase the risk of SAMs with high-dose atorvastatin, prompting consideration of an alternative statin (rosuvastatin) at a dose associated with lower SAMS risk.
- The patient's PGx results guided the optimization of their medication regimen, ensuring that they receive the most effective and safe treatment for their conditions.

Case 4

A 67-year-old African American female presents to the clinic with a complex past medical history significant for HTN, T2D, and ESRD on hemodialysis (MWF). Recently, she was diagnosed with a right lower extremity deep vein thrombosis (DVT) and was initiated on apixaban. However, due to financial constraints, she could not continue apixaban therapy and missed several doses, prompting her nephrologist to consider switching her to warfarin. The nephrologist enters a consult for assistance with warfarin initiation based on the patient's pharmacogenomic results.

Medical History:

ESRD on hemodialysis (MWF) (2 years)
Hypertension (25 years)
Type 2 diabetes (20 years)
Hyperlipidemia (17 years)
Chronic anemia (2 years)

Allergies:

None known

Social History:

Lives alone since the death of husband; assisted living in a continuing care retirement community; denies smoking or use of illicit substances.

Pertinent Vitals and Labs:

Height 5′6″, weight 132 lb, BP 130/80 mmHg, HR 70 BPM (non-dialysis days)
Complete blood count: Hgb 9 g/dL, Hct 28%, Plt 102, INR 1.1
Ferritin 222 ng/mL, TSat: 32%, Iron 20
Renal panel: Na 140 mmol, K+ 4.5 mmol, Cl- 100 mmol, BUN 30, SCr 4, PO4 5 mg/dL, corrected calcium 9 mg/dL, eGFR <15 mL/min, HbA1c 7%
Lipid panel: TG 145 mg/dL, TC 210 mg/dL, HDL 40 mg/dL, LDL 78 mg/dL
Liver panel: TBili 1 mg/dL, ALT 12 U/L, AST 20 U/L

Pharmacogenomic Results:

*CYP2D6 *1/*1* Normal metabolizer
*CYP2C19 *1/*1* Normal metabolizer
*SLCO1B1 *1/*1* Normal function
ABCG2 Normal function
*CYP2C9 *2/*5* Intermediate metabolizer
VKORC1 GA – Variant present
CYP4F2 CC – Normal activity
CYP2C cluster rs12777823 – Normal result

Current Medications:

Amlodipine 10 mg po daily
Atorvastatin 40 mg po daily
Calcitriol 0.25 mcg po daily
Furosemide 80 mg po daily
Metoprolol succinate XL 50 mg po daily
Sevelamer (Renvela) 800 mg po TID with meals

Medications Given during Hemodialysis:

Heparin IV 5000 units during hemodialysis
Iron sucrose (Venofer) IV 100 mg (total of 1 g or 10 doses) during hemodialysis
Erythropoietin (Procrit) IV 3000 units 3× a week during hemodialysis

Clinical (Nongenetic) Considerations

The primary concern for this patient is to achieve effective anticoagulation while minimizing the risk of bleeding, especially given her hemodialysis status [9]. Many direct oral anticoagulants (DOAC), aside from apixaban, are dependent on kidney function for elimination. Given that the patient's financial constraints are hindering her adherence to apixaban, warfarin may be a cost-effective option for this patient. Warfarin is highly protein-bound and unlikely to be dialyzed; HD patients tend to require about 20% lower doses compared to patients with normal renal function [35]. The use of warfarin will require INR monitoring, which can be done at the dialysis center, and allows the provider to monitor and make necessary dose adjustments.

Since the patient missed a few doses of apixaban (which explains her normal INR), bridging therapy will be required while warfarin dose is adjusted to achieve a therapeutic INR of 2–3. Options for bridging therapy include unfractionated heparin (UFH) and low-molecular-weight heparin (LMWH). Given that LMWH is renally eliminated, thus accumulating in patients with renal impairment, and is not cleared by dialysis, the use of enoxaparin in patients on HD is at provider discretion. At present, enoxaparin prescribing information does not give a specific dose recommendation for HD, rather a blanket recommendation of 1 mg/kg subcutaneously once daily for patients with a CrCl of less than 30 mL/min [14]. Additionally, the labeling speaks to an increased risk of hyperkalemia with enoxaparin in patients with severe renal dysfunction. Considering the patient is being seen in an ambulatory setting, the use of LMWH over UFH is preferred. A single-center study conducted by Pon et al. suggests therapeutic subcutaneous enoxaparin is safe for use at a reduced dose (0.7 mg/kg once daily) and anti-Xa monitoring (target 1–2 units/mL) may be used to ensure therapeutic dosing [30]. Of note, UFH or LMWH is typically administered during hemodialysis to prevent clotting of the dialyzer circuit. The dialysis center needs to be notified that patient is on therapeutic anticoagulation so appropriate adjustments can be made at the discretion of the nephrologist.

Pharmacogenomics Considerations

As the patient is of an African biogeographical ancestry group, it is most beneficial to confirm that the pharmacogenomics test panel included CYP2C9 alleles associated with patients of African biogeographical ancestry groups. In this case, the panel did include CYP2C9 alleles as it detected *5 allele.

Based on the patient's pharmacogenomic results, specifically her *CYP2C9* *2/*5 intermediate metabolizer phenotype and *VKORC1* GA genotype, a lower warfarin dose may be necessary to achieve the target INR range. The *VKORC1* GA genotype is associated with increased sensitivity to warfarin, requiring lower doses to avoid adverse effects. These genetic factors may affect the way the body processes warfarin, requiring a lower dose to maintain the optimal therapeutic effect [8]. The estimated maintenance warfarin dose that incorporates genetic and nongenetics can be determined using the warfarindosing.org calculator. To use this calculator:

1. Go to warfarindosing.org and select "0" for the number of warfarin doses taken.
2. Enter the patient's information into the warfarindosing.org online calculator.
3. Determine the estimated therapeutic warfarin dose, which is 2.2 mg po daily. However, this dose will be rounded to 2 mg po daily, and INR checks will be performed after 3 days to titrate the dose as needed.

The warfarindosing.org calculator does not include *CYP2C* cluster rs12777823, which is evaluated in patients with African ancestry since this variant was identified in patients of African biogeographical ancestry groups. This patient's *CYP2C* cluster rs12777823 result is normal, hence no dose adjustment based on *CYP2C* cluster is needed per CPIC guideline. This does not consider the patient's dietary vitamin K intake, which the pharmacist will have to counsel the patient on when providing her with warfarin education. Due to the patient's *CYP2C9* *2/*5 intermediate metabolizer phenotype, rapid warfarin dose adjustments should be avoided to prevent supratherapeutic INR levels.

Optimizing warfarin therapy requires careful consideration of the patient's dietary vitamin K intake, as it can significantly influence INR levels. To maintain stable INR levels, it is essential for the patient to consume a balanced diet with consistent vitamin K intake. Additionally, the patient's genetic predisposition, specifically her *VKORC1* GA genotype, may affect her response to vitamin K, a crucial nutrient for blood clotting and bone health. Leafy green vegetables, rich in vitamin K, may have a pronounced effect on her INR levels, and therefore, should be consumed consistently and in moderation to avoid fluctuations in anticoagulation.

Summary of Recommendations

- Due to financial constraints and the patient's ESRD status, transition from apixaban to warfarin is recommended.
- Start enoxaparin 1 mg/kg SQ once daily bridging while warfarin dosing is adjusted to achieve a therapeutic INR of 2–3.
- Use the warfarindosing.org calculator to estimate the patient's warfarin dose, considering her *CYP2C9* *2/*5 diplotype, indicating an intermediate metabolizer phenotype and *VKORC1 GA* genotype. The estimated maintenance warfarin dose is 2.2 mg po daily, rounded to 2 mg po daily.

- Educate the patient on maintaining a balanced diet with consistent vitamin K intake to optimize warfarin efficacy and safety.
- Regularly monitor the patient's INR levels, kidney function, and blood pressure, and adjust medications as necessary to ensure optimal management of her conditions.

What Did Pharmacogenomics Change About the Case Outcome?

- The patient's pharmacogenomic results, specifically her CYP2C9 intermediate metabolizer phenotype and *VKORC1 GA* genotype, influenced the warfarin dosing decision, leading to a lower than would typically be suggested for most patients estimated maintenance dose of 2 mg po daily.
- This personalized approach to warfarin therapy reduces the risk of supratherapeutic INR levels, potentially avoiding bleeding complications, and may help achieve faster time to goal INR compared to conventional dosing strategies, even those taking into consideration typical aging, polypharmacy, multimorbidity, and ancestry group considerations.
- By incorporating PGx results into the treatment plan, the healthcare team was able to tailor warfarin therapy to the patient's unique genetic profile, optimizing the balance between efficacy and safety.

References

1. 2023 American Geriatrics Society Beers Criteria® Update Expert Panel. American Geriatrics Society 2023 updated AGS Beers Criteria® for potentially inappropriate medication use in older adults. J Am Geriatr Soc. 2023;71(7):2052–81. https://doi.org/10.1111/jgs.18372.
2. Aïdoud A, Gana W, Poitau F, et al. High prevalence of geriatric conditions among older adults with cardiovascular disease. J Am Heart Assoc. 2023;12(2):e026850. https://doi.org/10.1161/JAHA.122.026850.
3. Alrajeh K, AlAzzeh O, Roman Y. The frequency of major *ABCG2*, *SLCO1B1* and *CYP2C9* variants in Asian, Native Hawaiian and Pacific Islander women subgroups: implications for personalized statins dosing. Pharmacogenomics. 2023;24(7):381–98. https://doi.org/10.2217/pgs-2023-0043. Epub 2023 May 24. PMID: 37222158; PMCID: PMC10242434.
4. Atkins JL, Delgado J, Pilling LC, Bowman K, Masoli JAH, Kuchel GA, Ferrucci L, Melzer D. Impact of Low Cardiovascular Risk Profiles on Geriatric Outcomes: Evidence From 421,000 Participants in Two Cohorts. J Gerontol A Biol Sci Med Sci. 2019;74(3):350-357. https://doi.org/10.1093/gerona/gly083.
5. Atorvastatin [package insert]. Memphis: Northstar Rx LLC; 2024. Accessed on 12 July 2025.
6. Cooper-DeHoff RM, Niemi M, Ramsey LB, et al. The clinical Pharmacogenetics implementation consortium guideline for SLCO1B1, ABCG2, and CYP2C9 genotypes and statin-associated musculoskeletal symptoms. Clin Pharmacol Ther. 2022;111(5):1007–21. https://doi.org/10.1002/cpt.2557.
7. Damluji AA, Forman DE, Wang TY, Chikwe J, Kunadian V, Rich MW, Young BA, Page RL 2nd, DeVon HA, Alexander KP; American Heart Association Cardiovascular Disease in Older Populations Committee of the Council on Clinical Cardiology and Council on Cardiovascular

and Stroke Nursing; Council on Cardiovascular Radiology and Intervention; and Council on Lifestyle and Cardiometabolic Health. Management of Acute Coronary Syndrome in the Older Adult Population: A Scientific Statement From the American Heart Association. Circulation. 2023;147(3):e32–e62. doi: 10.1161/CIR.0000000000001112. Epub 2022 Dec 12.

8. Dean L. Warfarin therapy and VKORC1 and CYP genotype. In: Pratt VM, Scott SA, Pirmohamed M, Esquivel B, Kattman BL, Malheiro AJ, editors. medical genetics summaries. Bethesda: National Center for Biotechnology Information (US); 2012.
9. Dimitrijevic ZM, Mitic BP, Tasic DD, Vrecic T, Paunovic K, Salinger S. Bleeding and thrombotic events in hemodialysis patients with atrial fibrillation on anticoagulation and antiplatelet therapy: a 24-month cohort study. Medicina (Kaunas). 2024;60(11):1760. https://doi.org/10.3390/medicina60111760.
10. Donato AJ, Machin DR, Lesniewski LA. Mechanisms of dysfunction in the aging vasculature and role in age-related Disease. Circ Res. 2018;123(7):825–48. https://doi.org/10.1161/CIRCRESAHA.118.312563.
11. Drenth-van Maanen AC, Wilting I, Jansen PAF. Prescribing medicines to older people-how to consider the impact of ageing on human organ and body functions. Br J Clin Pharmacol. 2020;86(10):1921–30. https://doi.org/10.1111/bcp.14094. Epub 2019 Dec 16. PMID: 31425638; PMCID: PMC7495267.
12. Duarte JD, Thomas CD, Lee CR, et al. Clinical Pharmacogenetics Implementation Consortium guideline (CPIC) for CYP2D6, ADRB1, ADRB2, ADRA2C, GRK4, and GRK5 genotypes and beta-blocker therapy. Clin Pharmacol Ther. 2024;116(4):939–47. https://doi.org/10.1002/cpt.3351. Epub 2024 Jul 1. PMID: 38951961; PMCID: PMC11502236.
13. Howlett SE, Fillit HM, Rockwood K, Woodhouse K, editors. Effects of aging on the cardiovascular system. In: Brocklehurst's textbook of geriatrics medicine and gerontology. Saunders Company; 2015.
14. Enoxaparin [package insert]. Bridgewater: Sanofi-Aventis U.S. LLC; 2022. Accessed on 25 July 2025.
15. Guasti L, Ambrosetti M, Ferrari M, et al. Management of hypertension in the elderly and frail patient. Drugs Aging. 2022;39(10):763–72. https://doi.org/10.1007/s40266-022-00966-7.
16. Johnson JA, Caudle KE, Gong L, et al. Clinical Pharmacogenetics Implementation Consortium (CPIC) guideline for pharmacogenetics-guided warfarin dosing: 2017 update. Clin Pharmacol Ther. 2017;102(3):397–404. https://doi.org/10.1002/cpt.668.
17. Kane M. Simvastatin therapy and SLCO1B1 genotype. In: Pratt VM, Scott SA, Pirmohamed M, et al., editors. Medical genetics summaries [Internet]. Bethesda: National Center for Biotechnology Information (US); 2024. Available from: https://www.ncbi.nlm.nih.gov/books/NBK602238/.
18. Kulkarni A, Mehta A, Yang E, Parapid B. Older adults and hypertension: beyond the 2017 guideline for prevention, detection, evaluation, and management of high blood pressure in adults. American College of Cardiology; 2020.
19. Lakatta EG, Levy D. Arterial and cardiac aging: major shareholders in cardiovascular disease enterprises: part II: the aging heart in health: links to heart disease. Circulation. 2003;107(2):346–54. https://doi.org/10.1161/01.cir.0000048893.62841.f7.
20. Rao SV, O'Donoghue ML, Ruel M, Rab T, Tamis-Holland JE, Alexander JH, Baber U, Baker H, Cohen MG, Cruz-Ruiz M, Davis LL, de Lemos JA, DeWald TA, Elgendy IY, Feldman DN, Goyal A, Isiadinso I, Menon V, Morrow DA, Mukherjee D, Platz E, Promes SB, Sandner S, Sandoval Y, Schunder R, Shah B, Stopyra JP, Talbot AW, Taub PR, Williams MS. 2025 ACC/AHA/ACEP/NAEMSP/SCAI Guideline for the Management of Patients With Acute Coronary Syndromes: A Report of the American College of Cardiology/American Heart Association Joint Committee on Clinical Practice Guidelines. Circulation. 2025;151(13):e771-e862. https://doi.org/10.1161/CIR.0000000000001309. Epub 2025 Feb 27. Erratum in: Circulation. 2025;151(13):e865. https://doi.org/10.1161/CIR.0000000000001328. Erratum in: Circulation. 2025;151(25):e1098. https://doi.org/10.1161/CIR.0000000000001346. Erratum in: Circulation. 2025;152(19):e402. https://doi.org/10.1161/CIR.0000000000001397. PMID: 40014670.

21. Lee CR, Luzum JA, Sangkuhl K, et al. Clinical Pharmacogenetics Implementation Consortium guideline for CYP2C19 genotype and clopidogrel therapy: 2022 update. Clin Pharmacol Ther. 2022;112(5):959–67. https://doi.org/10.1002/cpt.2526.
22. Jeon DS. Geriatric considerations in the management of elderly patients with cardiovascular diseases. Cardiovasc Prev Pharmacother. 2021;3(2):38–46. https://doi.org/10.36011/cpp.2021.3.e6.
23. Johnston SC, Amarenco P, Denison H, et al. Ticagrelor and aspirin or aspirin alone in acute ischemic stroke or TIA. N Engl J Med. 2020;383(3):207–17. https://doi.org/10.1056/NEJMoa1916870.
24. Clopidogrel [package insert]. Morristown: Sanofi-Aventis U.S. LLC; 2025. Accessed on 12 July 2025.
25. Kidney Disease: Improving Global Outcomes (KDIGO) CKD Work Group. KDIGO 2024 clinical practice guideline for the evaluation and management of chronic kidney disease. Kidney Int. 2024;105(4S):S117–314. https://doi.org/10.1016/j.kint.2023.10.018.
26. Mahaffey KW, et al. Ticagrelor compared with clopidogrel by geographic region in the Platelet Inhibition and Patient Outcomes (PLATO) trial. Circulation. 2011;124(5):544–54. https://doi.org/10.1161/CIRCULATIONAHA.111.047498.
27. Martin SS, Aday AW, Almarzooq ZI, et al. 2024 Heart disease and stroke statistics: a report of US and global data from the American Heart Association [published correction appears in Circulation. 2024;149(19):e1164. doi: 10.1161/CIR.0000000000001247]. Circulation. 2024;149(8):e347–913. https://doi.org/10.1161/CIR.0000000000001209.
28. Metoprolol succinate [package insert]. Bedminster: Ascend Laboratories, LLC; 2025. Accessed on 12 July 2025.
29. Pereira NL, Cresci S, Angiolillo DJ, et al. CYP2C19 genetic testing for oral P2Y12 inhibitor therapy: a scientific statement from the American Heart Association. Circulation. 2024;150(6):e129–50. https://doi.org/10.1161/CIR.0000000000001257.
30. Pon TK, Dager WE, Roberts AJ, White RH. Subcutaneous enoxaparin for therapeutic anticoagulation in hemodialysis patients. Thromb Res. 2014;133(6):1023–8. https://doi.org/10.1016/j.thromres.2014.03.036.
31. Powers WJ, Rabinstein AA, Ackerson, et al. Guidelines for the early management of patients with acute ischemic stroke: 2019 update to the 2018 guidelines for the early management of acute ischemic stroke: a guideline for healthcare professionals from the American Heart Association/American Stroke Association. Stroke. 2019;50:e344–418. https://doi.org/10.1161/STR.0000000000000211.
32. Ribeiro ASF, Zerolo BE, López-Espuela F, Sánchez R, Fernandes VS. Cardiac system during the aging process. Aging Dis. 2023;14(4):1105–22. https://doi.org/10.14336/AD.2023.0115.
33. Rolek B, Haber M, Gajewska M, Rogula S, Pietrasik A, Gąsecka A. SGLT2 inhibitors vs. GLP-1 agonists to treat the heart, the kidneys and the brain. J Cardiovasc Dev Dis. 2023;10(8):322. https://doi.org/10.3390/jcdd10080322.
34. Rosuvastatin [package insert]. Basking Ridge: Torrent Pharma Inc; 2025. Accessed on 12 July 2025.
35. Sakaan SA, Hudson JQ, Oliphant CS, et al. Evaluation of warfarin dose requirements in patients with chronic kidney disease and end-stage renal disease. Pharmacotherapy. 2014;34(7):695–702. https://doi.org/10.1002/phar.1445.
36. Simvastatin [package insert]. Somerset: Micro Labs USA, Inc; 2003. Accessed on 12 July 2025.
37. U.S. Department of Health and Human Services. Food and Drug Administration. Center for Drug Evaluation and Research (CDER). Pharmacokinetics in patients with impaired renal function – study design, data analysis, and impact on dosing. Guidance for industry. FDA; 2020. https://www.fda.gov/media/78573/download.
38. Vaduganathan M, Bhatt DL. Revisiting the clopidogrel-proton pump inhibitor interaction: from bench to bedside. Circ Cardiovasc Interv. 2015;8(10):e003208. https://doi.org/10.1161/CIRCINTERVENTIONS.115.003208. PMID: 26458412.

39. Wallentin L, Becker RC, Budaj A, et al. Ticagrelor versus clopidogrel in patients with acute coronary syndromes. N Engl J Med. 2009;361(11):1045–57. https://doi.org/10.1056/NEJMoa0904327.
40. Warden BA, Guyton JR, Kovacs AC, et al. Assessment and management of statin-associated muscle symptoms (SAMS): a clinical perspective from the National Lipid Association. J Clin Lipidol. 2023;17(1):19–39. https://doi.org/10.1016/j.jacl.2022.09.001. Epub 2022 Sep 11. PMID: 36115813.
41. Whelton PK, Carey RM, Aronow WS, et al. 2017 ACC/AHA/AAPA/ABC/ACPM/AGS/APhA/ASH/ASPC/NMA/PCNA guideline for the prevention, detection, evaluation, and management of high blood pressure in adults: executive summary: a report of the American College of Cardiology/American Heart Association Task Force on Clinical Practice Guidelines [published correction appears in Hypertension. 2018;71(6):e136–9. doi: 10.1161/HYP.0000000000000075]. [published correction appears in Hypertension. 2018;72(3):e33. doi: 10.1161/HYP.0000000000000080]. Hypertension. 2018;71(6):1269–324. https://doi.org/10.1161/HYP.0000000000000066.
42. Wiersinga J, Jansen S, Peters MJL, Rhodius-Meester HFM, Trappenburg MC, Claassen JAHR, Muller M. Hypertension and orthostatic hypotension in the elderly: a challenging balance. Lancet Reg Health Eur. 2024;48:101154. https://doi.org/10.1016/j.lanepe.2024.101154. PMID: 39717228; PMCID: PMC11665365.
43. Yazdanyar A, Newman AB. The burden of cardiovascular disease in the elderly: morbidity, mortality, and costs. Clin Geriatr Med. 2009;25(4):563–77, vii. https://doi.org/10.1016/j.cger.2009.07.007.
44. Zurek AM, Yendapally R, Urteaga EM. A review of the efficacy and safety of sodium-glucose cotransporter 2 inhibitors: a focus on diabetic ketoacidosis. Diabetes Spectr. 2017;30(2):137–42. https://doi.org/10.2337/ds16-0030.

Chapter 6
Pharmacogenomics and Pain Control for Older Adults

David Bright, Yee Ming Lee, Stephanie Sibicky, Youssef Roman, and Teresa DeLellis

Introduction

Pain management is a complex and challenging medical condition to diagnose and treat. In older adults, this complexity increases with the type and number of comorbidities, like osteoarthritis, chronic kidney disease, and cognitive impairment. Over 60% of residents in long-term care experience pain with a large proportion being underdiagnosed and undertreated [8]. In general, pain can be classified by duration (e.g., acute, chronic), intensity (e.g., mild, moderate, severe), and further by type of pain (e.g., neuropathic, inflammatory, cancer-associated) [7]. Acute pain often has an identifiable cause, like surgery or trauma, and is typically categorized as nociceptive (related to damage to body tissue). Untreated acute pain could contribute to the development of chronic pain, which can last months to years. However, the cause of chronic pain is not always known. Chronic pain often results from changes to the

D. Bright (✉)
Professor of Pharmaceutical Sciences, College of Pharmacy, Ferris State University, Big Rapids, MI, USA
e-mail: davidbright@ferris.edu

Y. M. Lee
Associate Professor of Clinical Pharmacy, Skaggs School of Pharmacy and Pharmaceutical Sciences, University of Colorado, Aurora, CO, USA

S. Sibicky
Clinical Professor and Assistant Dean for Academic Affairs and Assessment, Bouvé College of Health Sciences, Northeastern University, Boston, MA, USA

Y. Roman
L.S. Skaggs College of Pharmacy, Idaho State University, Pocatello, ID, USA

T. DeLellis
School of Pharmacy, College of Health Professions, Nursing and Pharmacy, Manchester University, Fort Wayne, IN, USA

T. DeLellis et al. (eds.), *Applying Pharmacogenomics to Optimize Geriatric Care*, https://doi.org/10.1007/978-3-032-10653-7_6

function of nerves and signal transmission, which may complicate treatment. Another type of pain related to nerve damage or abnormal functioning is neuropathic pain, often associated with conditions like fibromyalgia or diabetes [7].

Older adults may have differences in pain sensation compared to younger populations. While changes in central pain are unchanged, peripheral proprioception may be decreased [10]. Additionally, older adults experience different pharmacokinetic changes, such as an increase in body fat affecting volume of distribution, which can have significant implications on medication distribution. For example, lipid-soluble medications, such as fentanyl, may have an increased duration of action. A decrease in the mass of the liver and kidneys can also impact medication metabolism and clearance impacting the duration of action, potency, and elimination of medications warranting close monitoring for adverse effects and adjustments in dose based on hepatic or renal impairment (e.g., morphine, pregabalin) [1]. Pharmacodynamic changes, including alterations in receptor number and affinity as one ages, also contribute to an atypical response to medications [10, 12].

Recognizing the complexity of medication management for older adults, the American Geriatrics Society (AGS) Beers Criteria® includes an evaluation of medications for the management of pain in older adults [1]. They specifical recommend the avoidance of antidepressants with strong anticholinergic properties, like tricyclic antidepressants (e.g., amitriptyline), avoiding chronic use of NSAIDs due to an increased risk of gastrointestinal bleeding, and avoiding skeletal muscle relaxants (e.g., cyclobenzaprine) due to poor toleration of adverse effects. Additionally, NSAIDs and cyclooxygenase-2 inhibitors should be avoided in patients with heart failure due to the risk of fluid retention and heart failure exacerbations; TCAs in patients experiencing syncope; and opioids in patients with delirium or history of falls and fractures. Added sedation and risk of respiratory depression are also mentioned in combination, such as with multi-modal pain strategies of opioids and antidepressants, for example. Medications like gabapentin, tramadol, and baclofen are also included with warnings pertaining to their increased side effect potential in patients with a reduced glomerular filtration rate [1].

Despite the comprehensive considerations that are woven throughout the AGS Beers Criteria®, PGx is not overtly referenced as a factor leading to inclusion on the list. Therefore, it is important for pharmacists and other health care professionals practicing in the care of older adults to separately apply consideration of PGx factors when making clinical geriatric recommendations. Below, three cases begin to describe how PGx factors can be woven together with different types of pain to create a comprehensive approach to pain management in older adults.

Pharmacogenomics (PGx) provides additional information to be considered in combination with clinical factors such as drug–drug interactions, drug–disease interactions, and pharmacokinetic and pharmacodynamic changes that accompany aging physiology. These genetic and clinical factors can be used to guide medication selection and dosing in the therapeutic management of pain for older adults. There are pharmacogenomic guidelines available for certain nonsteroidal anti-inflammatory drugs (NSAIDs), opioids and pain-modulating antidepressant medications (see Tables 6.1, 6.2, and 6.3).

Table 6.1 Nonsteroidal anti-inflammatory drugs and CYP2C9 [14]

CYP2C9 phenotype (activity score; AS)		CPIC recommendations
CYP2C9 normal metabolizer		*Short-acting NSAID: celecoxib, flurbiprofen, lornoxicam, and ibuprofen* *and intermediate-acting: meloxicam* *and long-acting NSAID: piroxicam and tenoxicam* Initiate therapy with recommended starting dose. Use the lowest effective dosage for the shortest duration consistent with individual patient treatment goals.
CYP2C9 intermediate metabolizer	(AS 1.5)	*Short-acting NSAID: celecoxib, flurbiprofen, ibuprofen* *and intermediate-acting: meloxicam* *and long-acting NSAID: piroxicam and tenoxicam* Initiate therapy with recommended starting dose. Use the lowest effective dosage for the shortest duration consistent with individual patient treatment goals.
	(AS 1.0)	*Short-acting NSAID: celecoxib, flurbiprofen, lornoxicam, and ibuprofen* Initiate therapy with the lowest recommended starting dose. Titrate dose upward to clinical effect or maximum recommended dose with caution. *Intermediate-acting: meloxicam* Initiate therapy with 50% of the lowest recommended starting dose. Titrate dose upward to clinical effect or 50% of the maximum recommended dose with caution. Use the lowest effective dosage for shortest duration consistent with individual patient treatment goals. Upward dose titration should not occur until after steady-state is reached (at least 7 days). *Long-acting NSAID: piroxicam and tenoxicam* Use an alternative therapy not metabolized by CYP2C9 or shorter-acting NSAID.
CYP2C9 poor metabolizer		*Short-acting NSAID: celecoxib, flurbiprofen, lornoxicam, and ibuprofen* Initiate therapy with 25–50% of the lowest recommended starting dose. Titrate dose upward to clinical effect or 25–50% of the maximum recommended dose with caution. Use the lowest effective dosage for the shortest duration consistent with individual patient treatment goals. Upward dose titration should not occur until after steady-state is reached (at least 8 days for celecoxib and 5 days for ibuprofen, flurbiprofen, and lornoxicam after the first dose in PMs). *Intermediate-acting: meloxicam* Use an alternative therapy not metabolized by CYP2C9 or shorter-acting NSAID. *Long-acting NSAID: piroxicam and tenoxicam* Use an alternative therapy not metabolized by CYP2C9 or shorter-acting NSAID.

Table 6.2 Opioids and CYP2D6 [4]

CYP2D6 phenotype	CPIC recommendations
CYP2D6 ultrarapid metabolizer	Codeine and tramadol: Avoid using codeine or tramadol because of potential for serious toxicity. Hydrocodone: No recommendation due to limited data available.
CYP2D6 normal metabolizer	Codeine, hydrocodone, and tramadol: Use label recommended dose.
CYP2D6 intermediate metabolizer	Codeine, hydrocodone, and tramadol: Use label recommended dose. If no response and opioid use is warranted, consider non-codeine and non-tramadol opioid[a].
CYP2D6 poor metabolizer	Codeine and tramadol: Avoid using codeine or tramadol due to risk of diminished analgesia. If opioid use is warranted, consider non-codeine and non-tramadol opioid[a]. Hydrocodone: Decreased metabolism of hydrocodone to its active metabolite, hydromorphone, but data are limited to determine if this pharmacokinetic effect translates into altered analgesia or side effects. Use label recommended dose. If no response and opioid use is warranted, consider non-codeine and non-tramadol opioid[a].

[a]Opioids not significantly impacted by CYP2D6 genetic polymorphism include fentanyl, hydromorphone, methadone, morphine, and oxymorphone

Table 6.3 Venlafaxine and CYP2D6 [3]

CYP2D6 phenotype	CPIC recommendations
CYP2D6 ultrarapid metabolizer	No recommendation due to limited data.
CYP2D6 normal metabolizer	Initiate therapy with recommended starting dose.
CYP2D6 intermediate metabolizer	No recommendation due to limited data.
CYP2D6 poor metabolizer	The clinical impact of altered venlafaxine metabolite to venlafaxine ratio is unclear but poor metabolizer genotype has been associated with adverse effects. Consider a clinically appropriate alternative not predominantly metabolized via CYP2D6.

Pharmacogenomics of NSAIDs: The Role of CYP2C9

Nonsteroidal anti-inflammatory drugs remain one of the most widely used classes of medications for managing pain and inflammation in conditions such as gout, rheumatoid arthritis, and osteoarthritis. While effective, their use is associated with important safety concerns, including gastrointestinal bleeding, hypertension, and kidney injury. These risks are compounded by the fact that NSAIDs exhibit wide variability in half-life and pharmacokinetics, making dosing and monitoring essential to balance therapeutic benefit with safety.

Pharmacogenomics provides a framework to better understand and predict individual variability in NSAID response. The metabolism of many NSAIDs, particularly diclofenac, celecoxib, meloxicam, piroxicam, flurbiprofen, and ibuprofen, occurs primarily through CYP2C9, a highly polymorphic enzyme. Common reduced-function alleles, such as *2 and *3, can decrease enzymatic activity, leading to higher systemic exposure and greater risk of adverse effects, including gastrointestinal bleeding. In contrast, some NSAIDs, such as naproxen, ketorolac, and aspirin, rely less on CYP2C9 metabolism and may be less affected by genetic variability in this pathway. Incorporating pharmacogenomic information into prescribing decisions could therefore allow clinicians to tailor NSAID selection and dosing to patient-specific genetic profiles, optimizing efficacy while minimizing toxicity [14].

Pharmacogenomics of Opioids: The Role of CYP2D6

Opioids are an important class of medications for the management of acute and breakthrough pain. Although long-term use is generally discouraged due to risks of dependence and adverse effects, opioids remain a valuable option for patients who do not achieve adequate pain relief with standard therapies or for whom NSAIDs are contraindicated [13].

Despite their effectiveness, clinical responses to opioids vary widely across individuals. A key reason for this variability is genetic differences in drug-metabolizing enzymes. Several opioids, including codeine and tramadol, are prodrugs that require metabolic activation to exert their analgesic effects. This conversion is primarily mediated by the enzyme CYP2D6, one of the most polymorphic pharmacogenes in the CYP450 enzymes. Genetic variants in *CYP2D6* result in phenotypes ranging from poor metabolizers, who may experience little or no pain relief, to ultrarapid metabolizers, who face an elevated risk of toxicity due to excessive formation of active metabolites. To address these differences, current clinical PGx guidelines provide clear recommendations for avoiding codeine and tramadol in patients identified as poor or ultrarapid metabolizers, with the use of alternative analgesics preferred. This highlights the growing role of PGx in tailoring opioid therapy to optimize both safety and efficacy [4].

Case 1

A 65-year-old female with no significant past medical history underwent a total hip replacement after she sustained a fracture from a fall during her walk in the snow. Postoperatively, her surgeon prescribed the following medications, but she complained of pain (score 7/10) that limited her participation in physical rehabilitation.

Pertinent Vitals and Labs:

5′1″, 132 lb., BMI 25, BP 125/80, HR 75.
Post-op labs: Hgb 14 g/dL, platelets normal, SCr 0.8 (CrCl = 66 ml/min).

Pharmacogenomic Results:

*CYP2C9 *1/*3* (activity score 1) Intermediate metabolizer.
*CYP2C19 *1/*2* Intermediate metabolizer.
*CYP2D6 *1/*1* Normal metabolizer.

Drug allergies: None.
Current Medications:

Enoxaparin 40 mg subcutaneously once daily.
Acetaminophen 500 mg by mouth Q6H PRN.
Tramadol 50 mg by mouth Q8H PRN.

Clinical (Nongenetic) Considerations

The treatment goal for this patient is to optimize pain management so she can participate in physical therapy to facilitate mobilization, retain her quality of life and independence, and prevent debilitation. Current utilization of the "as needed" acetaminophen and tramadol orders are key pieces of information to collect to inform decision-making. Because the hip fracture was caused by a fall, the use of opioids such as tramadol, which can cause sedation and increase falls risk (as discussed above from the AGS Beers Criteria®) should be minimized or avoided when possible. Renal function should also be assessed prior to treatment decisions to inform medication selection and dosing. This patient does not present with multimorbidity or polypharmacy, which simplifies treatment decisions and poses minimal risk for phenoconversion when interpreting her pharmacogene phenotypes. Depending on the information collected, options for this patient's pain regimen include nonpharmacologic modalities, adjusting acetaminophen, short-term use of NSAIDs, adjusting the tramadol dose, or changing to another opioid medication.

The first option is to schedule acetaminophen (e.g., 1000 mg every 8 hours). While unlikely to achieve pain and function goals alone, it is a safe and effective means of supporting pain relief in older adults who are not heavy users of alcohol and without liver disease. This approach can facilitate less exposure overall to other higher-risk medications.

Another option is to add a short course of an NSAID to potentiate the effects of tramadol in addition to scheduling acetaminophen. Because she does not have known cardiovascular disease or major cardiovascular risk factors, bleeding risk would be the only concern to mitigate, especially considering enoxaparin use. If this patient is deemed appropriate to continue enoxaparin, bleeding risk may be more thoroughly assessed first by checking for a history of gastritis and obtaining a hemoglobin level. NSAID use should be limited to a short course and then reassessed. If chronic NSAID use is chosen, a proton pump inhibitor (PPI) could be considered to prevent GI bleeding and ulcers. This approach would need to be weighed against the risk of bringing this patient closer to polypharmacy and considering that chronic PPI therapy is also impacted by her CYP2C19 phenotype.

A third option would be the modification of the tramadol order. A dose could be scheduled for at least 30 minutes prior to each therapy session. Depending on the patient's response, a higher dose could also be ordered. Adverse effects such as sedation, dizziness, and constipation should be closely monitored. If constipation occurs but the patient is responding well in all other respects, a stimulant laxative (e.g., sennosides) may be initiated until the tramadol dose can be decreased as healing progresses.

Finally, changing tramadol to another opioid could be considered if her pain remains uncontrolled. Hydrocodone would be a reasonable option if this route is chosen since it is a short-acting opioid. Because of the risks discussed above of opioid therapy, this option should be reserved in case of tramadol failure [1].

Pharmacogenomic Considerations

There are no known PGx parameters that affect acetaminophen use. In choosing a specific NSAID, certain NSAIDs (see Table 6.1) are significantly impacted by CYP2C9 metabolism. This patient has CYP2C9 intermediate metabolizer (IM) phenotype (activity score 1). CPIC guidelines have separate recommendations for short, intermediate, and long-acting NSAIDs based on CYP2C9 phenotype. For short-acting NSAIDs (celecoxib, ibuprofen), CPIC recommends initiating the lowest effective dose for the shortest duration consistent with patient treatment goals. For intermediate-acting NSAIDs (meloxicam), CPIC recommends a 50% dose reduction in initial dose and waiting at least seven days before increasing the dose. CPIC recommends avoiding long-acting NSAIDs (e.g., piroxicam) due to an increased risk of toxicity from drug accumulation. Another option is to use an NSAID not impacted by CYP2C9 such as naproxen.

Because the patient exhibits a CYP2D6 normal metabolizer phenotype, a standard older adult dose can be used according to CPIC guidelines, and a standard response should be expected following standard dosing. In effect, the PGx data suggest that the patient's genetics are not the root cause of the insufficiency of the pain management approach. However, particularly in the case of the older adult, this PGx-based finding does not negate the need for close monitoring for and addressing of adverse effects as discussed above. Hydrocodone is less impacted by CYP2D6 than tramadol (see Table 6.2), but as noted above, should be reserved for cases of tramadol failure.

Summary of Recommendations

- Shared decision-making and further patient assessment should be utilized to determine the choice between making acetaminophen scheduled, adding an NSAID, increasing the tramadol dose or frequency, or switching to another opioid.
- Any treatment plan should include a multimodal approach including non-pharmacologic therapy and should be closely monitored for opportunities to deescalate.
- Piroxicam and opioids should be avoided. Tramadol, celecoxib, or ibuprofen are the safest options for this scenario, with naproxen reserving special consideration given the lack of adjustment due to CYP2C9 phenotype, and therapy should be limited to the lowest effective dose for the shortest time possible with judicious monitoring to avoid accumulation and evaluate for or minimize the need for additional treatment with laxatives or PPIs.

What Did Pharmacogenomics Change About the Case Outcome?

- Pharmacogenomics guidance will not direct analgesic class choice in this case; other clinical factors are more appropriate in this scenario. However, PGx did confirm the relative safety of tramadol and other opioids, meaning that a standard dose of an opioid should provide a standard response.
- If an NSAID is chosen, PGx data may help to inform NSAID selection and dosing so as to minimize adverse effects and maximize efficacy.

Case 2

A 76-year-old male presents with an acute gout flare, his second episode in three months, following one episode every few years since age 70. He attempted dietary changes after his first gout flare, but he states that "it's really hard" to adhere to dietary recommendations, and he appears unwilling to be strongly adherent with lifestyle adjustments. Prior treatment has consisted of NSAIDs and/or colchicine, depending on the treating physician, but urate-lowering therapy has not been previously attempted.

Medical History:

Diabetes
Hypertension
Dyslipidemia
Gout

Pertinent Vitals and Labs:

SCr 1.0 (CrCl = 85–90 ml/min)
Height: 6′2”
Weight: 240 lbs

Pharmacogenomic Results:

*HLA-B*58:01*: Negative; Noncarrier
*CYP2C9 *3/*3* Poor metabolizer
Ancestry group: European

Current Medications:

Metformin 1000 mg by mouth BID.
Lisinopril 10 mg by mouth daily.
Atorvastatin 80 mg by mouth daily.

Clinical (Nongenetic) Considerations

This patient presents a common scenario of acute gout flares that increase with age, presumably with lifestyle contributions. Not all patients are willing to undergo deeper lifestyle adjustments, and there is growing evidence that dietary changes have less impact on outcomes than medications [9, 11]. There is even some indication about a genetic predisposition to diet affecting gout flares, although testing for this predisposition is not available [18]. Considering all these factors, gout flares may increase in frequency over time. Following a track record of gout flares for years, and especially with a second gout flare in a 3-month window, this patient qualifies for urate-lowering treatment (ULT) per the 2020 American College of Rheumatology (ACR) Gout guidelines [6]. These guidelines also recommend the initiation of concomitant anti-inflammatory prophylaxis for 3–6 months during the initiation of ULT. Options to consider are corticosteroids, NSAIDs, and colchicine. Beyond ULT, there are drug–disease interactions that can be optimized within the context of multimorbidity.

ACR guidelines drive a preference of allopurinol for gout flare management. It may be reasonable to start allopurinol 100 mg daily for this patient, considering there appear to be minimal aging- or disease-related changes in renal function and no drug–drug interactions with his current medication list. Monitoring should consist of serum urate (SU) measurements to achieve a target of <6 mg/dL to reduce the risk of subsequent gout flares.

Anti-inflammatory options each pose various risks, particularly with older adults, that need to be considered in the context of potential benefit. Of the three options listed, corticosteroids carry the most risk for the patient considering their existing diabetes and hypertension. NSAIDs can also disrupt hypertension control, increase risk of AKI (especially in older adults and in combination with ACE inhibitors), and increase risk of GI bleeding [1]. While this patient does not have impaired renal function or other GI bleeding risks at this time, they do have an indication for low-dose aspirin for cardiovascular event prevention based on their age and risk equivalents. If NSAIDs are chosen for short-term therapy, initiation of low-dose aspirin could be delayed until that course is complete. Colchicine is generally considered to be a safe option, particularly with CrCl >30 mL/min. In patients over 70 years of age, it may be prudent to reduce the dose by 50%, which coincides with ACR recommendations for lower colchicine dosing over the 3–6-month window of colchicine use during allopurinol initiation and titration. The potential drug–drug interaction with atorvastatin can increase risk of myopathy, further supporting the 50% dose reduction if initiating colchicine. GI adverse effects would need to be closely monitored, especially for diarrhea, which could lead to dehydration, increasing the risk of acute kidney injury with his antihypertensive therapy. Low-dose colchicine has also recently achieved FDA approval for use in reducing cardiovascular events.

Beyond gout-specific treatment, there are several opportunities to maximize medication therapy benefits while reducing related risks. While many antihypertensives can precipitate gout flares, losartan has the unique benefit of lowering uric acid [17]. Changing lisinopril to losartan would preserve the renal protective effects for patients with both hypertension and diabetes, while having a positive effect on gout. Depending on what the patient's hemoglobin A1c is (at or above goal), there are several diabetes medications with beneficial drug–disease interactions in this scenario. There is growing evidence to support the efficacy of SGLT-2 inhibitors in reducing the incidence of gout and can also have beneficial effects outside of blood glucose lowering, including reduction in cardiovascular event risk, delay of renal disease progression, and mortality benefits in chronic heart failure [15]. While this patient does not have renal disease or heart failure, they do have risk factors for both. Blood glucose and urogenital infections would have to be monitored more closely as euglycemic diabetic ketoacidosis and infection risk increase with age [1]. GLP-1 agonists do not have adverse effects on gout, also reduce the risk of cardiovascular events, and may help this patient achieve weight loss goals.

Pharmacogenomic Considerations

Allopurinol is a medication with an adverse event PGx interaction potential. Pharmacogenomic guidelines indicate that patients who are *HLA-B*58:01* allele carriers have an increased risk for allopurinol-induced severe cutaneous adverse reaction (SCAR), among other cutaneous reactions. This patient has tested negative for the *HLA-B*58:01* biomarker, meaning that there is a low risk of SCAR and no reason to prescribe an alternate therapy to allopurinol. Of note, ACR guidelines recommend *HLA-B*58:01* testing in patient of Southeast Asian or African biogeographic ancestry groups but recommend *against* such PGx testing in other populations [6]. Therefore, if this patient did not already have PGx testing on file, it would *not* be recommended to order a PGx test specifically for *HLA-B*58:01* prior to allopurinol initiation at this time. It is worth noting that many PGx tests are offered as part of a larger panel; therefore, if PGx testing for other uses was warranted, *HLA-B*58:01* results may be included in the report even if not specifically clinically indicated. Therefore, having *HLA-B*58:01* data in the patient's records is not necessarily an indication of inappropriate laboratory testing, but is more likely the result of the patient previously obtaining a broad panel of PGx data in prior laboratory testing.

PGx testing can assist with stratifying some of the inherent risks with anti-inflammatory use in older adults. From a PGx standpoint, certain NSAIDs should receive specific dosing adjustments based on this patient's *CYP2C9* poor metabolizer phenotype. CPIC NSAID guidelines recommend using significantly reduced doses of short-acting NSAIDs, such as celecoxib and ibuprofen. However, other NSAIDs, such as naproxen, are less impacted by *CYP2C9* genetic polymorphism and would be acceptable alternatives from a PGx perspective.

Summary of Recommendations

- Allopurinol is indicated and does not have significant safety concerns at standard starting doses for this patient.
- For the first 3–6 months of allopurinol therapy, low-dose colchicine or an NSAID appropriate to her phenotype (see Table 6.1), while ensuring adequate hydration, are appropriate options to be discussed in the shared decision-making process.
- In addition to gout-specific therapy, changing lisinopril to losartan, initiating low-dose aspirin, and initiating either an SGLT-2 inhibitor or GLP-1 agonist if current hemoglobin A1c allows, are all options that should be discussed with the patient to optimize overall medication therapy.

What Did Pharmacogenomics Change About the Case Outcome?

- Negative *HLA-B*58:01* results did not change the case outcome, and the PGx test would not have been appropriate to order if PGx testing had not been done.
- CYP2C9 poor metabolizer phenotype was highly pertinent for NSAID selection and dosing, with naproxen or colchicine (low dose) as appropriate alternatives to ibuprofen.
- Panel-based testing is often more cost-effective, including *CYP2C9* with *HLA* testing provided additional information to inform drug prescribing.

Case 3

AJ is a 72-year-old male who presents with persistent and debilitating pain on the right thoracic region following a recent episode of shingles. His skin is severely blistered.

Medical History:

Diabetes with neuropathy
Hypertension
Benign prostatic hypertrophy
Anxiety
Depression
Dyslipidemia

Pharmacogenomic results:

*CYP2C19 *1/*2* Intermediate metabolizer
*CYP2D6 *1/*4* Intermediate metabolizer
*HLA-B*15:02* Positive

Current Medications:

Metformin 1000 mg by mouth BID
Semaglutide 7 mg by mouth daily
Gabapentin 100 mg by mouth TID
Terazosin 10 mg by mouth qHS
Nifedipine XL 60 mg by mouth daily
St. John's Wort daily
Lisinopril 10 mg by mouth daily
Atorvastatin 40 mg by mouth daily

Clinical (Nongenetic) Considerations

Treating neuropathic pain in older adults is a common clinical conundrum, since essentially all medications indicated appear at some level in the AGS Beers Criteria [1]. Choice of first-line agents depends on the underlying pathology causing the pain (e.g., diabetic neuropathy, postherpetic neuralgia). Agents with the strongest recommendations across categories include low-dose amitriptyline and other tricyclic antidepressants (TCAs), gabapentin or pregabalin, duloxetine or other serotonin-norepinephrine reuptake inhibitors (SNRIs), anticonvulsants (e.g., carbamazepine, oxcarbazepine), and topical therapies (e.g., lidocaine, capsaicin) [5].

TCAs have the most evidence and are often the most efficacious. They also carry the highest risks due to strong anticholinergic effects (e.g., dementia, delirium, falls, orthostasis) [1]. Because this patient is already on two medications with high risk of orthostasis (terazosin, nifedipine), TCAs would be particularly dangerous. They should therefore be reserved for cases when the patient's goals weigh neuropathic pain control above long-term risks and other therapies have failed to achieve either pain control or tolerability goals.

Topical therapies are the safest option due to minimal systemic exposure and thus adverse effects, drug–drug, and drug–disease interactions [2]. They are a reasonable first option if the patient's skin is intact and it is not applied over sensitive areas such as the face or genitals. However, they may be less efficacious in deeper than subcutaneous pain, and there have been minimal studies in this area. Because this patient has severe blistering, topical therapy may not be appropriate due to the risk of systemic absorption.

Duloxetine and other SNRIs can be efficacious options if topical therapies are unsuitable or ineffective, but carry some treatment risks especially in older adults [1]. SNRIs are appropriate options especially if co-morbidities such as anxiety, depression, or stress incontinence are present. Renal function should be assessed first because dose reductions are necessary in CrCl <30 mL/min. For this patient with concurrent anxiety and depression, his anxiety and depression scores should also be assessed. If anxiety and depression control are not optimal, SNRIs would be a sensible neuropathic pain treatment option. This would interact with the serotonergic properties of St. John's Wort, so this supplement would need to be discontinued. Additional monitoring is necessary in older adults for sodium levels at initiation and with any dose changes due to the increased risk of the syndrome of inappropriate antidiuretic hormone (SIADH), which is more common in older adults. SNRIs may worsen uncontrolled hypertension; hence, close blood pressure monitoring may be warranted. Additional caution should be exercised in older adults with a fall or fracture history, which this patient does not have.

Gabapentin and pregabalin are first-line options for postherpetic neuralgia [5]. Dose reductions are indicated if CrCl <60 mL/min, so renal function should be assessed before use and periodically during use along with monitoring for sedation and other CNS effects. Because this patient is already on low-dose gabapentin for neuropathy, increasing the dose to relieve postherpetic neuralgia pain is reasonable if the patient is not already experiencing adverse effects and is on two or fewer other medications that have CNS depressant effects. In contrast to the other treatment options discussed, this approach would not contribute further to polypharmacy for this patient.

The anti-epileptic medications, carbamazepine and oxcarbazepine, are primarily indicated for trigeminal neuralgia, but are at times used off-label if all other options have been exhausted. They carry the same sodium monitoring needs in older adults as SNRIs, with frequent drug–drug interactions because they are both metabolized by and strong inducers of CYP3A4. For this patient, the interaction with St. John's Wort and atorvastatin would be problematic.

If none of the above neuropathic pain options are efficacious after a sufficient trial, short-term tramadol or opioids can be trialed in an attempt to preserve quality of life, with the limitations discussed with Case 1 above.

Pharmacogenomic Considerations

CPIC guidelines offer recommendations for TCAs based on CYP2D6 and CYP2C19 phenotypes, but the data are based on antidepressant doses studied, which are higher than those used for neuropathic pain. This patient has a CYP2D6 and CYP2C19 intermediate metabolizer phenotype for many TCAs, which may predispose him to medication accumulation. Considering the safety limitations for TCAs in older adults in general, pharmacogenomics is unlikely to have a great impact except to serve as a further data point toward the safety risks to inform shared decision-making conversations if needed.

Pharmacogenomics data for lidocaine and capsaicin are currently limited and require further research. No actionable pharmacogenomics data exists for gabapentin or pregabalin. There is a CPIC guideline available for venlafaxine but for this patient who has a CYP2D6 intermediate metabolizer phenotype, guidelines recommend following standard dosing, confirming that a typical dose is likely to provide a typical response (see Table 6.3). Before using carbamazepine or oxcarbazepine, CPIC guidelines provide recommendations for assessing risk of severe cutaneous adverse reactions based on HLA result. This patient is *HLA-B*15:02* allele positive, which contraindicates the use of these two medications.

Summary of Recommendations

- The most reasonable option depends on response to current therapy for comorbid conditions.
- If they are not experiencing adverse effects to the current gabapentin dose, this could be increased. Increasing the dose of gabapentin would not further polypharmacy, but a follow-up should be scheduled to ensure a gradual dose reduction is attempted after the postherpetic neuralgia has improved.
- If anxiety or depression are not optimally controlled, the addition of an SNRI would be reasonable. If SNRIs are used, sodium should be monitored closely, and St. John's wort should be discontinued.
- Options to avoid include topical therapies because patient's skin is not intact and anti-epileptic medications due to the *HLA-B*15:02* result associated with increased risk for severe cutaneous adverse reaction, the lack of evidence in this type of neuralgia, and drug–drug interactions.
- If gabapentin therapy is not successful, TCAs may be used with caution as a last remaining option with close monitoring based on cognitive and orthostatic risks.

What Did Pharmacogenomics Change About the Case Outcome?

- While PGx results were relevant to the selection of treatment options, they ultimately did not change the outcome for this case because gabapentin was ultimately chosen for neuropathic pain management, and gabapentin dose adjustments are not necessary based on the PGx data provided.
- *HLA-B*15:02* is positive for risk allele associated with carbamazepine and oxcarbazepine induced severe cutaneous adverse reactions. Carbamazepine was already ruled out as an enzyme inducer causing risk for drug–drug interactions and because of alternative treatment options with strong evidence for this condition.
- Considering the clinical limitations of efficacy and safety of TCAs in older adults, PGx testing is unlikely to be impactful in this context except as a further data point toward safety risks to inform shared decision-making conversations if needed and affordable.

References

1. 2023 American Geriatrics Society Beers Criteria® Update Expert Panel. American Geriatrics Society 2023 updated AGS beers criteria® for potentially inappropriate medication use in older adults. J Am Geriatr Soc. 2023;71(7):2052–81. https://doi.org/10.1111/jgs.18372. Epub 2023 May 4. PMID: 37139824
2. American Geriatrics Society Beers Criteria® Alternatives Panel, Steinman MA, Semla TP, Akgün KM, et al. Alternative treatments to selected medications in the 2023 American Geriatrics Society Beers Criteria®. J Am Geriatr Soc. 2025:1–21. https://doi.org/10.1111/jgs.19500.
3. Bousman CA, Stevenson JM, Ramsey LB, et al. Clinical Pharmacogenetics Implementation Consortium (CPIC) Guideline for CYP2D6, CYP2C19, CYP2B6, SLC6A4, and HTR2A genotypes and serotonin reuptake inhibitor antidepressants. Clin Pharmacol Ther. 2023;114(1):51–68. https://doi.org/10.1002/cpt.2903. Epub 2023 May 30. PMID: 37032427; PMCID: PMC10564324
4. Crews KR, Monte AA, Huddart R, et al. Clinical Pharmacogenetics implementation consortium guideline for CYP2D6, OPRM1, and COMT genotypes and select opioid therapy. Clin Pharmacol Ther. 2021;110(4):888–96. https://doi.org/10.1002/cpt.2149. Epub 2021 Feb 9. PMID: 33387367; PMCID: PMC8249478
5. Cruccu G, Truini A. A review of neuropathic pain: from guidelines to clinical practice. Pain Ther. 2017;6(Suppl 1):35–42. https://doi.org/10.1007/s40122-017-0087-0. Epub 2017 Nov 24. PMID: 29178033; PMCID: PMC5701894
6. FitzGerald JD, Dalbeth N, Mikuls T, Brignardello-Petersen R, Guyatt G, Abeles AM, Gelber AC, Harrold LR, Khanna D, King C, Levy G, Libbey C, Mount D, Pillinger MH, Rosenthal A, Singh JA, Sims JE, Smith BJ, Wenger NS, Bae SS, Danve A, Khanna PP, Kim SC, Lenert A, Poon S, Qasim A, Sehra ST, Sharma TSK, Toprover M, Turgunbaev M, Zeng L, Zhang MA, Turner AS, Neogi T. 2020 American College of Rheumatology Guideline for the Management of Gout. Arthritis Care Res (Hoboken). 2020;72(6):744–60. https://doi.org/10.1002/acr.24180. Epub 2020 May 11. PMID: 32391934; PMCID: PMC10563586

7. Herndon CM, Kominek CM, Mullins AM. Pain Management. In: DiPiro JT, Yee GC, Haines ST, Nolin TD, Ellingrod VL, Posey L, editors. DiPiro's pharmacotherapy: a pathophysiologic approach. 12th ed. McGraw Hill; 2023. Accessed May 02, 2024. https://accesspharmacy-mhmedical-com.ezproxy.neu.edu/content.aspx?bookid=3097&sectionid=266106683.
8. Knopp-Sihota JA, MacGregor T, Reeves JTH, Kennedy M, Saleem A. Management of Chronic Pain in long-term care: a systematic review and meta-analysis. J Am Med Dir Assoc. 2022;23(9):1507–1516.e0. https://doi.org/10.1016/j.jamda.2022.04.008. Epub 2022 May 18. PMID: 35594944
9. Major TJ, Topless RK, Dalbeth N, Merriman TR. Evaluation of the diet wide contribution to serum urate levels: meta-analysis of population based cohorts. BMJ. 2018;363:k3951. https://doi.org/10.1136/bmj.k3951. PMID: 30305269; PMCID: PMC6174725
10. Rajan J, Behrends M. Acute pain in older adults: recommendations for assessment and treatment. Anesthesiol Clin. 2019;37(3):507–20. https://doi.org/10.1016/j.anclin.2019.04.009. Epub 2019 Jul 1. PMID: 31337481
11. Roman YM. Moving the needle in gout management: the role of culture, diet, genetics, and personalized patient care practices. Nutrients. 2022;14(17):3590. https://doi.org/10.3390/nu14173590. PMID: 36079846; PMCID: PMC9460297
12. Schwan J, Sclafani J, Tawfik VL. Chronic pain Management in the Elderly. Anesthesiol Clin. 2019;37(3):547–60. https://doi.org/10.1016/j.anclin.2019.04.012. Epub 2019 Jun 18. PMID: 31337484; PMCID: PMC6658091
13. Sturgeon IM, Roman YM. Individualized Management of Osteoarthritis: the role of pharmacogenomics to optimize pain therapy. Future Pharmacol. 2025;5(2):30.
14. Theken KN, Lee CR, Gong L, et al. Clinical Pharmacogenetics implementation consortium guideline (CPIC) for CYP2C9 and nonsteroidal anti-inflammatory drugs. Clin Pharmacol Ther. 2020;108(2):191–200. https://doi.org/10.1002/cpt.1830. Epub 2020 Apr 28. PMID: 32189324; PMCID: PMC8080882
15. Wei J, Choi HK, Dalbeth N, et al. Gout flares and mortality after sodium-glucose Cotransporter-2 inhibitor treatment for gout and type 2 diabetes. JAMA Netw Open. 2023;6(8):e2330885. https://doi.org/10.1001/jamanetworkopen.2023.30885. PMID: 37624597; PMCID: PMC10457713
16. Wolff ML, Cruz JL, Vanderman AJ, Brown JN. The effect of angiotensin II receptor blockers on hyperuricemia. Ther Adv Chronic Dis. 2015;6(6):339–46. https://doi.org/10.1177/2040622315596119. PMID: 26568810; PMCID: PMC4622312
17. Zhang Y, Yang R, Dove A, et al. Healthy lifestyle counteracts the risk effect of genetic factors on incident gout: a large population-based longitudinal study. BMC Med. 2022;20(1):138. https://doi.org/10.1186/s12916-022-02341-0. PMID: 35484537; PMCID: PMC9052486

Chapter 7
Pharmacogenomics and Central Nervous System Care for Older Adults

Nadia Khartabil, Yee Ming Lee, David Bright, Stephanie Sibicky, and Teresa DeLellis

Abbreviations

AD	Alzheimer's disease
AED	Antiepileptic drug
AGS	American Geriatrics Society
Beers criteria	AGS beers criteria® for potentially inappropriate medication use in older adults
BID	Twice daily (bis in die)
BP	Blood pressure
CBZ	Carbamazepine
CNS	Central nervous system
COMT	Catechol-O-Methyltransferase
CrCl	Creatinine clearance

N. Khartabil (✉)
School of Pharmacy, West Coast University, Los Angeles, CA, USA
e-mail: NKhartabil@westcoastuniversity.edu

Y. M. Lee
Associate Professor of Clinical Pharmacy, Skaggs School of Pharmacy and Pharmaceutical Sciences, University of Colorado, Aurora, CO, USA

D. Bright
Professor of Pharmaceutical Sciences, College of Pharmacy, Ferris State University, Big Rapids, MI, USA

S. Sibicky
Clinical Professor and Assistant Dean for Academic Affairs and Assessment, Bouvé College of Health Sciences, Northeastern University, Boston, MA, USA

T. DeLellis
School of Pharmacy, College of Health Professions, Nursing and Pharmacy, Manchester University, Fort Wayne, IN, USA

T. DeLellis et al. (eds.), *Applying Pharmacogenomics to Optimize Geriatric Care*, https://doi.org/10.1007/978-3-032-10653-7_7

CYP	Cytochrome P450
CYP2B6	Cytochrome P450 2B6
CYP2C9	Cytochrome P450 2C9
CYP2C19	Cytochrome P450 2C19
ECG	Electrocardiogram
eGFR	Estimated glomerular filtration rate
FRAT score	Falls risk assessment tool score
GERD	Gastroesophageal reflux disease
GI	Gastrointestinal
HLA-B	Human leukocyte antigen B (gene; variants such as *HLA-B*15:02 are associated with drug hypersensitivity)
HR	Heart rate
LTG	Lamotrigine
MAO-B	Monoamine oxidase B
NSAID	Nonsteroidal anti-inflammatory drug
PD	Parkinson's disease
PGx	Pharmacogenomics
PO	Per Os (by mouth)
prn	Pro re Nata (as needed)
QT	QT interval
q8h	Every 8 hours
SSRI	Selective serotonin reuptake inhibitor

Introduction

Neurological diseases, particularly those affecting the central nervous system (CNS), are among the most challenging conditions to treat, owing to their complex pathophysiology and interplay of genetic, environmental, and age-related factors. Older age is a well-established risk factor for many CNS disorders, including Alzheimer's disease, Parkinson's disease, stroke, and epilepsy, as age-related changes in brain structure, vascular integrity, and cellular repair mechanisms exacerbate vulnerability to these conditions [13]. Furthermore, the Global Burden of Disease Study highlights that neurological conditions remain the leading contributor to disability-adjusted life years (DALYs) worldwide, underlining the urgent need for innovative therapeutic strategies [14].

Pharmacogenomics (PGx) has emerged as a promising tool in CNS pharmacotherapy, offering the potential to tailor treatments based on an individual's genetic profile with the hope of reducing side effects and improving efficacy. Key PGx markers with clinically actionable information and PGx guidelines for CNS medications include *CYP2C19, CYP2B6,* and *CYP2D6*, which influence the metabolism of several antidepressants and antipsychotics, and *HLA* gene, which is associated with severe hypersensitivity reactions to certain antiepileptic drugs. There are additional PGx biomarkers studied for other CNS medications, but the data are currently inconclusive regarding their clinical utility. Some of these biomarkers may be included on commercial PGx panels with no PGx guideline to provide clinical recommendations. Hence, providers may highlight for patients why certain results cannot be used due to limited evidence. Examples of these PGx biomarkers include *SLC6A4*, associated with serotonin transporter functionality, and *COMT*, which impacts dopamine metabolism [5].

Despite the clinical utility of PGx in CNS pharmacotherapy, it is important to recognize that PGx provides only one piece of a larger and more complex puzzle to consider, and it is imperative that PGx considerations not distract from the patient's other clinical factors. Polypharmacy, altered pharmacokinetics, and pharmacodynamics associated with aging complicate treatment regimens, as do age-related genetic and epigenetic changes that influence drug response. Additionally, certain CNS drugs may have poor blood–brain barrier penetration and/or narrow therapeutic indices that are further exacerbated by comorbidities common in older populations [16].

This chapter examines the intersection of CNS pharmacotherapy and PGx in older adults, highlighting how genetic variations affect drug efficacy and safety in this population. This discussion aims to advance personalized medicine approaches to CNS diseases, improving outcomes in an aging population burdened by neurological disorders, and highlighting the limitations in translating genetic data into actionable interventions.

Pharmacogenomics of Antidepressants and Antipsychotics: The Role of **CYP2C19, CYP2B6,** *and* **CYP2D6**

See Chap. 4 for pharmacogenomics associated with antidepressants and antipsychotics.

Pharmacogenomics of Antiepileptic Drugs: The Role of **HLA and CYP2C9**

Genetic variation in human leukocyte antigen (HLA) is associated with an increased risk of developing cutaneous adverse reactions to certain antiepileptic drugs. The variant allele HLA-B**15:02* is strongly associated with an increased risk of Stevens–Johnson syndrome (SJS) and toxic epidermal necrolysis (TEN) in patients treated with carbamazepine, oxcarbazepine, and phenytoin. The variant allele *HLA-A**31:01 is associated with a greater risk of maculopapular exanthema, drug reaction with eosinophilia and systemic symptoms, and SJS/TEN in patients treated with carbamazepine.

CPIC guidelines provide antiepileptic recommendations based on HLA-B and HLA-A genotypes (see Table 7.1). Additionally, the FDA labeling for

Table 7.1 Antiepileptic drugs gene–drug interaction guideline [17, 19, 23]

HLA genotype	CPIC and DPWG recommendations
HLA-B*15:02 POSITIVE and any HLA-A*31:01 genotype	CPIC guideline: For new starts, avoid using carbamazepine, oxcarbazepine, and phenytoin. For patients who have been taking carbamazepine, oxcarbazepine, or phenytoin for at least 3 months without any cutaneous adverse effects, cautiously consider the use of carbamazepine, oxcarbazepine, or phenytoin and monitor for adverse effects. Although the evidence linking other aromatic antiepileptic drugs (such as esclicarbazepine, lamotrigine, phenobarbital) and cutaneous adverse reaction with HLA-B*15:02 is weak, caution should be used in choosing an alternative drug. DPWG guideline: Avoid using lamotrigine if an alternative is available. Consider the risk of SJS/TEN against the benefits of using lamotrigine. If lamotrigine is used, advise the patient to report any skin rash immediately.
HLA-B*15:02 negative HLA-A*31:01 POSITIVE	Carbamazepine: For new starts, avoid using carbamazepine. For patients who have been taking carbamazepine for at least 3 months without any cutaneous adverse effect, cautiously consider use of carbamazepine and monitor for adverse effects. As the evidence linking other aromatic antiepileptic drugs and cutaneous adverse reaction with HLA-A*31:01 is very limited, CPIC currently has no recommendation regarding the selection of other aromatic antiepileptic drugs. Previous tolerance to carbamazepine is not indicative of tolerance to other aromatic antiepileptic drugs.
HLA-B*15:02 negative HLA-A*31:01 negative	Use carbamazepine, oxcarbazepine and phenytoin per standard dosing guidelines and monitor for adverse effects.

carbamazepine includes a boxed warning to avoid using carbamazepine in patients tested positive for the HLA-B*15:02 variant unless the benefit outweighs the risk. The label highlights that the HLA-B*15:02 variant is found almost exclusively in patients with ancestry across broad areas of Asia, hence it requires HLA-B*15:02 testing prior to initiating carbamazepine in patient with "ancestry in genetically at-risk populations." In patients testing positive for HLA-B*15:02 variant, there may be a risk of cross-reactivity with other aromatic antiepileptic drugs including eslicarbazepine, lamotrigine, and phenobarbital.

Phenytoin is metabolized by hepatic CYP2C9. Genetic polymorphisms of CYP2C9 result in decreased CYP2C9 activity, thereby increasing phenytoin levels. CPIC guidelines provide CYP2C9-guided phenytoin maintenance dose recommendations in patients who are HLA-B*15:02 negative (see Table 7.2). Of note, this CYP2C9-guided dosing information is useful for patients newly started on phenytoin as subsequent dose adjustments are based on therapeutic drug monitoring.

Table 7.2 CPIC dosing recommendation for phenytoin based on HLA-B*15:02 and CYP2C9 phenotype [17, 19, 23]

HLA-B*15:02 genotype and CYP2C9 phenotype	CPIC recommendation
HLA-B*15:02 negative[a] and CYP2C9 normal or intermediate metabolizer (activity score 1.5)	Use typical dosing strategies, and adjust subsequent doses based on therapeutic drug monitoring, response, and side effects reported.
CYP2C9 intermediate metabolizer (activity score 1)	Use typical initial or loading dose. For subsequent doses, reduce typical maintenance dose by 25%. Adjust subsequent doses based on therapeutic drug monitoring, response, and side effects.
CYP2C9 poor metabolizer	Use typical initial or loading dose. For subsequent doses, reduce typical maintenance dose by 50%. Adjust subsequent doses based on therapeutic drug monitoring, response, and side effects.

[a]HLA-B*15:02 negative result does not eliminate the risk of phenytoin-induced SJS/TEN. Monitor for cutaneous adverse effects

Case 1

An 86-year-old male residing in a skilled facility has encountered several recent medication-related challenges in epilepsy management. He was previously stable on phenytoin for over 20 years. Over the last several years as his dementia has progressed, he has become frail, sarcopenic, and facility medical and support staff have struggled to maintain adequate nutritional status. This led to several recent hospitalizations related to fluctuating phenytoin levels, and he transitioned to levetiracetam at his last discharge. In the ensuing months, aggressive behavior has become a progressive problem. One of his providers at the facility ordered a PGx panel to assist in identifying alternate pharmacotherapy.

Medical History:

Epilepsy
Alzheimer's dementia
Essential tremor
Depression
Osteoarthritis
Asthma

Allergies:

None known

Pertinent Vitals and Labs:

HR 60 bpm, BP 110/60, Albumin 2.5 mg/dL, CrCl 60–65 mL/min

Pharmacogenomic Results:

*HLA-B*15:02* Positive
*CYP2B6 *1/*1* normal metabolizer
*CYP2C19 *17/*17* ultrarapid metabolizer
*CYP2C9 *1/*1* normal metabolizer

Current Medications:

Memantine/donepezil 28 mg/10 mg PO daily
Levetiracetam 500 mg PO BID
Propranolol 80 mg PO daily
Citalopram 20 mg PO daily
Acetaminophen 1000 mg PO q8h
Meloxicam 7.5 mg PO BID prn for pain
Budesonide/formoterol 160/4.5, 2 puffs BID
Budesonide/formoterol 160/4.5, 1 puff prn for difficulty breathing (notify medical team with each use)
Polyethylene glycol 17 gm PO daily prn 3 days without bowel movement

Clinical (Nongenetic) Considerations

The primary concern to address in this case is dementia with behavioral challenges. While there are many potential optimizations to address in this case, this discussion will focus on this primary concern, touching on additional conditions and treatment plans as they intersect. The first and most important step to approaching behavioral challenges is to assess and address multifactorial contributing factors [7]. In this case, some possibilities that should be explored are asthma control, changes in diet or bowel habits, pain control, mood, and potentially exacerbating medications [12].

Mood is an area where many factors intersect, including aging physiology, polypharmacy, multimorbidity, and PGx. See Chap. 4 for pharmacogenomics associated with antidepressants and antipsychotics. An evaluation of depression should be undertaken. While the Geriatric Depression Scale (GDS) has validity evidence in people with cognitive impairment, a true assessment of symptoms in advanced dementia may still pose a challenge [27]. Additionally, citalopram carries a dose-related risk of prolonged QT syndrome and arrhythmias. Propranolol, formoterol, and donepezil also independently increase of arrhythmias, so if depression treatment is insufficient an alternate treatment may be necessary.

Pain control is another complicated area. Behavior changes in advanced dementia could be a nonverbal sign of pain, and as needed pain medications are frequently underutilized in patients who cannot respond to typical pain scales [2]. It would be reasonable to give a dose of meloxicam and observe for resolution of behavior challenges. However, when the extent of his aging and comorbidities is taken into account, if scheduled pain medications are indicated, a topical therapy such diclofenac or other alternate therapy should be considered to minimize his already high risk of gastrointestinal bleeding ([4] updated AGS Beers Criteria® for potentially inappropriate medication use in older adults [1]).

Perhaps the most challenging to address in this case is the reasonable possibility that levetiracetam could be the genesis of aggressive behavior. All alternative agents for epilepsy have complicated risk and benefit scenarios based on his phenotype, aging, and multimorbidity. Therefore, the first step would be to assess the time since his last seizure to determine if deprescription of antiepileptics would be reasonable. The case notes recent hospitalizations from phenytoin fluctuations, likely related to low albumin from his poor nutritional status, but does not specify if these were due to adverse effects or lack of efficacy (i.e., seizures). If an antiepileptic is still warranted or desired by the patient's family, carbamazepine (CBZ), lamotrigine (LTG), or gabapentin are reasonable options to evaluate.

CBZ has some preliminary evidence to suggest it may be helpful in refractory symptoms of agitation or aggression in patients with dementia at doses of 300–600 mg per day [21]. Dosing would need to be carefully monitored and adjusted for hypoalbuminemia (target 1–2 mcg/mL trough), and sodium closely monitored based on the increased risk of SIADH when used in older adults in general [18]. One reason to avoid CBZ would be the risk of exacerbating tremors. The tremor concerns for this patient and exacerbating and remitting factors are discussed in more detail below as relates to gabapentin use.

Lamotrigine can be effective for some forms of epilepsy, and is better tolerated than phenytoin or CBZ in some patients [10]. Lamotrigine also has potentially mood-stabilizing effects, which could benefit this patient. However, this patient's nutritional status can hinder absorption, making it difficult to achieve target concentrations. In addition, lamotrigine can exacerbate essential tremors, which is problematic in this case for reasons discussed below under gabapentin.

The final alternative medication to consider for epilepsy would be gabapentin. Depending on the type and history of seizures, this could help address both seizures and the essential tremor. This could also work synergistically with nonpharmacologic interventions for agitation, along with discontinuing the levetiracetam, if pain is an agitating factor. Finally, a unique benefit to gabapentin could be a consolidation of medications. His original epilepsy treatment of phenytoin could have worsened essential tremors, which could have led to the propranolol prescription. Propranolol could worsen asthma control by interacting with his formoterol inhaler. Difficulty breathing could contribute to agitation and to increased inhaler use, which could in turn further exacerbate tremors. Second, propranolol can have psychiatric adverse effects itself, potentially worsening depression and contributing to agitation. This patient's current heart rate and blood pressure are low, increasing the risk of falls and syncopal events, especially when combined with acetylcholinesterase inhibitors and citalopram. Therefore, gabapentin could help with epilepsy, tremor, and agitation, especially if its use facilitates deprescribing of propranolol. It is important to note, however, that gabapentin itself can increase the risk of falls.

Pharmacogenomic Considerations

This patient has a *CYP2C19* ultra-rapid metabolizer phenotype, and therefore likely to receive insufficient effect from citalopram due to low drug levels. It is uncertain how much dose increment is necessary to overcome this rapid metabolism to achieve therapeutic levels with the risk of exceeding the FDA label-recommended maximum dose of 20 mg per day in older adults (citalopram [product insert]. North Chicago, IL: AbbVie, Inc.; 2023. Accessed July 21, 2025). Polypharmacy and multimorbidity further increase risk in this case as noted above. To ensure his depression is well controlled and not exacerbating behavior challenges, it would be reasonable to change to another SSRI such as sertraline. Sertraline does not increase QT prolongation. Sertraline is metabolized by multiple pathways including CYP2C19 and CYP2B6. Current CPIC guidelines do not require a dose adjustment for patients with a CYP2C19 ultrarapid metabolizer phenotype, as there is only a small increase in sertraline metabolism compared to CYP2C19 normal metabolizer phenotypes. Therefore, sertraline can be initiated at the otherwise standard dose based on his CYP2C19 and CYP2B6 phenotypes [6].

While CBZ and lamotrigine are options with pros and cons to weigh from a non-genetic clinical standpoint, this patient's PGx results eliminate them both as options. HLA-B*15:02 positive patients are at increased risk of severe cutaneous hypersensitivity reaction, including SJS/TENs (Table 7.1). Despite his tolerance of phenytoin in the past, the risk associated with CBZ in this phenotype contraindicates that treatment option for this patient. There have been reports of lamotrigine-induced SJS/TENs even though the risk is 0.4% [19]. DPWG guidelines recommend avoiding lamotrigine if an alternative drug is available (Table 7.1).

There is no PGx guidance available for gabapentin. There is PGx information for meloxicam based on CYP2C9. Given his CYP2C9 normal metabolizer phenotype and lack of renal disease or hypertension, the current meloxicam dose is safe [26].

Summary of Recommendations

Consider the time since the last seizure. This could be an opportunity to deprescribe his antiepileptic.

If antiepileptic therapy is necessary, effects on mood, agitation, falls risk, and tremors should be considered. CBZ and lamotrigine should be avoided due to increased risk for severe hypersensitivity reactions associated with *HLA-B*15:02* variant.

Based on his PGx results, gabapentin could be the safest option and may also help deprescribe propranolol, which could exacerbate several conditions while increasing falls and arrhythmia risk. Based on recent agitation, PGx phenotype, and drug–drug and drug–disease interactions, it would be reasonable to change citalopram to sertraline.

Regardless of pharmacotherapy choices, non-pharmacologic assessments and modalities should be employed to help ease agitation and improve nutrition.

What Did PGx Change About the Case Outcome?

PGx assisted in evaluating epilepsy treatment options, where clinical geriatrics presented a mix of pros and cons. There was no clear best option in terms of risk versus benefit for advanced aging physiology, due to conflicting drug–disease and drug–drug interactions, as well as multimorbidity and polypharmacy.

PGx suggested that the patient's citalopram was likely subtherapeutic due to *CYP2C19* ultrarapid metabolizer phenotype, helping add some clarity to the complicated evaluation of depression control. A change in therapy could be advised to better help with mood control and potentially agitation, rather than further increasing citalopram dose, which would exceed the FDA label-recommended dose in older adults.

Case 2

You are completing a scheduled chart review on an 84-year-old male nursing home resident. Upon interviewing the patient, he expresses wanting to remove as many medications as possible because he is "tired of all the pills" and "just wants to live the rest of my life in peace." Nursing staff also point out that this patient has had difficulty with bowel movements over the past month or so since the increase in solifenacin dose due to multiple incontinent episodes. You confirm with the patient's family and providers that this goal is reasonable and begin the process of eliminating unnecessary medications and deprescribing.

Medical History:

Alzheimer's dementia
Mixed urinary incontinence (urge, overflow)
Hypertension
GERD
Iron-deficiency anemia
Benign prostatic hyperplasia (BPH)
Osteoporosis s/p fall five years ago

Allergies

None known

Pertinent Vitals and Labs:

BP 134/68 (range 112–136/68–80 in last 60 days), CrCl = 42 mL/min

Pharmacogenomic Results:

CYP2D6 *1/*5 (intermediate metabolizer)

Current Medications:

Donepezil 10 mg daily
Lisinopril 10 mg PO daily
Solifenacin 10 mg PO daily
Finasteride 5 mg PO daily
Tamsulosin 0.4 mg PO daily
Pantoprazole 40 mg PO daily
Calcium carbonate 500 mg PO every 6 hours PRN indigestion
Ferrous sulfate 325 mg PO every other day
Acetaminophen 1000 mg PO three times a day
Hydromorphone 1 mg PO every 8 hours prn pain
Polyethylene glycol 17 gm PO twice daily
Senna 1 tablet PO daily

Clinical (Nongenetic) Considerations

Preparing for deprescribing and eliminating unnecessary medications includes thoroughly reviewing the patient's medical conditions and medications. Recognizing the immediate drug interaction between an acetylcholinesterase inhibitor (donepezil) and an anticholinergic (solifenacin) should raise concerns—especially since the patient is experiencing increased incontinence episodes and bouts of constipation warranting additional treatment. This is an example of the prescribing cascade and leads to an increased risk of medication-related problems due to polypharmacy. While mechanisms are counter to each other, close monitoring of symptoms can be conducted to ensure that the anticholinergic medication is preventing urge incontinence episodes with adequate effect. While the patient has had more incontinence episodes, medications may not be the most effective treatment strategy as evidence notes a reduction in incontinence episodes by 0.5 episodes per day. (Hsu) This is also complicated by the patient's BPH and overflow incontinence. Anticholinergic medications increase the risk of urinary retention, which can worsen overflow incontinence. Adding other medications that can be constipating (e.g., opioids, iron, calcium) can also exacerbate constipation and overflow incontinence. Other pharmacodynamic interactions between medications that increase the risk of orthostatic hypotension (tamsulosin, lisinopril) should be considered.

As medications are eliminated or deprescribed, it is important to consider that enhanced adverse effects and potential new adverse events are possible, so close monitoring should be employed.

Based on the drug–drug interaction between donepezil and solifenacin and the unfavorable efficacy vs adverse effect profile bladder antimuscarinics, solifenacin should be considered for adjustment or deprescribing. If solifenacin is deprescribed after discussion with the patient and care partners, a slow approach to tapering can be utilized if breakthrough symptoms occur [11]. If an agent for incontinence is still desired and within the patient's treatment goals, alternatives include adding a beta-3 agonist, like mirabegron 25 mg daily and titrating to 50 mg, if tolerated. Multiple considerations need to be made before adding this therapy, including kidney function, control of the patient's hypertension (agent should be avoided in severe, uncontrolled hypertension, defined as blood pressure $\geq$ 180/110). Vibergron is another beta-3 agonist option, although use be limited in patients with reduced kidney function (eGFR <15 mL/min/1.73m^2) or uncontrolled hypertension.

A conversation about the long-term treatment of BPH and this patient's life expectancy should also take place when the patient and caregivers are ready, as prolonged use of alpha-blockers should be reassessed frequently [22]. Medications that can increase the risk of constipation (hydromorphone, calcium carbonate) should also be evaluated for continued use as constipation can contribute to urinary incontinence and impact quality of life. Iron supplementation should be evaluated as a potential cause of constipation, and checking iron studies for this patient may reveal that iron is no longer needed.

Pharmacogenomic Considerations

The patient is currently taking donepezil, which is a minor substrate of CYP2D6. Some researchers recommend a dose adjustment by 25–50% in *CYP2D6* intermediate metabolizer phenotype due to an increased exposure to donepezil, and a higher likelihood of experiencing side effects (e.g., incontinence) [8]. Adjusting the patient's dose from 10 mg daily to 5 mg daily may be appropriate to maintain the efficacy of the agent and reduce the risk of adverse effects. An alternative acetylcholinesterase inhibitor, like rivastigmine, which is not known to be metabolized through the CYP450 system, could mitigate some of these concerns.

After adjusting the dose of donepezil or considering an alternative, the team should discuss the discontinuation of solifenacin. The adverse effects of higher concentrations of donepezil may contribute to incontinence episodes and the prescribing cascade described above. If the beta-3 agonist alternative agents are explored, PGx consideration of the CYP2D6 metabolism status and mirabegron's potential to cause drug–drug interactions as a moderate CYP2D6 inhibitor adds to the clinical picture. Altered CYP2D6 metabolism may have limited impact on mirabegon's overall elimination as in vitro studies indicate *CYP2D6* poor metabolizer phenotype had a 16% higher drug concentration than normal metabolizer phenotype [20]. Mirabegron is a moderate CYP2D6 inhibitor that can inhibit the already reduced CYP2D6 function (intermediate metabolizer) and reduce the metabolism of donepezil and tamsulosin, making it a less favorable option. Vibegron does not have the same CYP2D6 interactions.

PGx considerations can also add to the discussion regarding long-term use of tamsulosin as mentioned above. Similar to donepezil, tamsulosin concentrations can be increased in a patient who has a *CYP2D6* intermediate metabolizer phenotype, potentially leading to a higher risk of side effects including orthostatic hypotension, dizziness, and falls, which are concerning for this patient. While this is a mechanism-based consideration, specific guidance about managing this interaction is not currently available for CYP2D6 intermediate metabolizer phenotypes, as there is no CPIC guideline except a recommendation in the FDA label to use lower doses in CYP2D6 poor metabolizer phenotypes [25]. It would be reasonable to review the dose of tamsulosin based on patient's clinical symptoms.

Summary of Recommendations

- Discuss with patient and care partners decreasing donepezil to 5 mg.
- Reinforce non-pharmacogenomic therapy for incontinence; monitor closely for improvement when/if donepezil is decreased and discuss deprescribing solifenacin with patient and care partners to improve cognition and reduce risk of exacerbating BPH. Mirabegron or vibegron is a reasonable alternative if mediation therapy for incontinence is desired.
- When/if solifenacin is successfully deprescribed, consider deprescribing tamsulosin to reduce falls risk with close monitoring for BPH symptoms.
- Re-evaluate need for and dosing of hydrocodone, iron, and calcium (including potential multimodal alternatives) in an effort to reduce constipation risk and the resulting exacerbation of incontinence and BPH symptoms.

What Did PGx Change About the Case Outcome?

- PGx results provided valuable insight into the likelihood and potential resolution strategies for prescribing cascade in this patient. The patient having a *CYP2D6* intermediate metabolizer phenotype could be leading to donepezil accumulation, which can result in increased drug efficacy. However, there is a need to monitor for adverse effects, including incontinence, that may prompt dose adjustments.
- PGx results provide additional information to justify the elimination of unnecessary medications, like solifenacin, and seeking alternative agents if necessary.
- PGx provided some clarity to the otherwise nebulous discussion regarding potential adjustments to other medications like tamsulosin. Due to increased concentration in patients who have *CYP2D6* intermediate metabolizer phenotypes, there

is a higher risk of adverse effects like hypotension, dizziness, and falls, making dose reduction trials a more concrete recommendation.

Case 3

A 71-year-old male with a 10-year history of Parkinson's disease (PD) presents with worsening motor symptoms. The patient is frustrated as the increase in his carbidopa/levodopa caused episodes of dyskinesia. Now with the lower dosing, he feels like he can't even enjoy walking around his backyard without tripping. He also reported difficulty with fine motor tasks such as writing and buttoning his shirt. He fears losing independence and is feeling isolated due to these limitations. His wife is his main caregiver who took early retirement to take care of her husband. The patient's wife asks if there is any PGx testing that can help manage his medications better.

Medical History:

Constipation
Depression
Increased risk of falls, 1 recent fall with no fractures
Orthostatic hypotension
Osteoarthritis
Osteoporosis
Idiopathic Parkinson's disease
Rapid eye movement sleep disorder (RBD)

Allergies:

None known

Pertinent Vitals and Labs:

BP 102/65 mm Hg
HR 70 bpm
Timed up and Go (TUG): 20 seconds
Hoehn and Yahr scale: 3
FRAT score: 17

Pharmacogenomic Results:

*CYP2D6 *1/*4* Intermediate metabolizer
*CYP2C19 *1/*1* normal metabolizer

Current Medications:

Acetaminophen 650 mg q 8 hours prn
Alendronate 70 mg weekly
Calcium/vitamin D 500 mg /400 IU 1 tablet twice daily

Docusate 100 mg twice daily as needed
Paroxetine 20 mg daily
Carbidopa/levodopa 100/25 mg 1 tablet three times a day
Rasagiline 0.5 mg daily
Opicapone 50 mg at bedtime
Amantadine 100 mg three times a day
Benztropine 0.5 mg three times a day
Droxidopa 100 mg three times a day
Melatonin 5 mg at bedtime

Clinical (Nongenetic) Considerations

In this patient with idiopathic Parkinson's disease (PD) and worsening motor symptoms, treatment typically focuses on optimizing dopaminergic therapies to manage motor fluctuations. The current regimen includes multiple agents such as carbidopa/levodopa, rasagiline, opicapone, and amantadine, all aimed at stabilizing dopamine levels and reducing symptoms like bradykinesia, rigidity, and tremors. The dyskinesia experienced by the patient suggests a need for careful dose adjustments to avoid peaks in dopamine levels that can trigger these motor complications.

The patient's concurrent conditions (constipation, orthostatic hypotension, depression, falls) necessitate adjustments that address non-motor symptoms and minimize adverse effects from medications, particularly those with anticholinergic properties.

There are several problems to assess that may have contributed to the patient's worsening of symptoms. Treatment-induced motor fluctuations are a complex multifactorial problem that the patient encounters for several possible reasons. Motor complications may result from disease progression, drug-induced dopamine fluctuation, and drug-induced dopamine deficiency.

There are several nongenetic factors influencing therapeutic decision-making. First is an increased sensitivity to medications. As PD progresses, patients tend to become more sensitive to fluctuation in dopamine levels, leading to motor complications. Second, both PD and its medications can exacerbate orthostatic hypotension. This patient's blood pressure of 102/65 mmHg raises concern, and further monitoring or treatment modifications may be necessary. This patient is on droxidopa for orthostatic hypotension. MAO-B inhibitors do not appear to affect droxidopa clearance. Carbidopa may reduce droxidopa clearance, although it has not been determined to be clinically significant [3]. Third, anticholinergic medications such as benztropine can contribute to cognitive decline, urinary retention, slowed gastric motility, and worsening constipation in older adults [1]. Benztropine in combination with paroxetine can contribute to additional anticholinergic adverse effects. This slowed gastrointestinal (GI) motility is particularly relevant for the absorption of carbidopa/levodopa, the cornerstone of PD treatment. Delayed gastric emptying may reduce the drug's bioavailability, leading to delayed or unpredictable onset of its therapeutic effects, which could contribute to the patient's "off periods" and worsening motor symptoms. This reduced and erratic absorption could also partially explain the patient's difficulties with maintaining adequate symptom control despite adjustments in levodopa dosing. The final factor to consider is this patient's risk of falls. His FRAT score (17) indicates a high fall risk. The high risk of falls is compounded by multiple factors such as orthostatic hypotension, polypharmacy, impaired motor function, and the use of anti-cholinergic medications.

Optimizing dopaminergic therapy can be improved by addition of extended-release carbidopa/levodopa or increasing rasagiline. Extended-release carbidopa/levodopa addition may offer a smoother dopaminergic stimulation to improve gait and reduce "off" episodes, yet it may pose a risk of dyskinesia. Increasing the dose of rasagiline is another option to improve motor function. At 1 mg daily, rasagiline has shown greater efficacy in reducing "off" time and enhancing motor performance when used as an adjunct to levodopa therapy. According to the PRESTO and LARGO trials, it was demonstrated that rasagiline 1 mg reduced daily "off" time by 0.9–1.2 hours when added to levodopa [24]. Increasing the dose of rasagiline, however, may exacerbate hypotension in combination with levopdopa. Blood pressure monitoring is essential.

The patient is already on opicapone, the once daily catechol-O-methyltransferase (COMT) inhibitor, which extends the effect of levodopa by reducing its peripheral metabolism. Clinical trials such as BIPARK I and II have demonstrated that opicapone can significantly reduce "off episodes and improve motor symptoms control". However, in real-world settings, its efficacy can be compromised by suboptimal adherence. In this case, where the patient is reporting worsening motor function, it is essential to evaluate adherence to opicapone and the overall medication regimen before making dose changes or therapeutic additions [15].

Addressing overlapping pharmacological and non-motor complications (such as GI motility, depression, orthostatic hypotension) as a whole may have a significant impact on medication efficacy and quality of life. Anticholinergic burden is higher in this patient; discontinuation of benztropine may lower the burden contributing to falls, confusion, and constipation as mentioned above.

Pharmacogenomics Considerations

The patient's *CYP2D6* intermediate metabolizer phenotype (*1/*4) impacts his metabolism of paroxetine, leading to higher drug exposure and an increased risk of anticholinergic adverse effects. These include worsened constipation, orthostatic hypotension, increased fall risk, and cognitive impairment, all of which are further complicating this patient's PD and symptom management as described above. Paroxetine's anticholinergic burden can exacerbate the patient's already present constipation and contribute to delayed gastric emptying, a common non-motor symptom in PD patients. Therefore, changing paroxetine to another SSRI with reduced potential for anticholinergic adverse effects should be considered. Given his *CYP2C19* normal metabolizer phenotype, an SSRI metabolized via CYP2C19 such as escitalopram may be preferred for this patient. Refer to Chap. 4 for details on PGx and antidepressants for SSRI selection.

As for Parkinson's disease, one of the PGx biomarkers frequently investigated is the genetic polymorphism in dopaminergic receptor (e.g., DRD2), which has been associated with L-dopa-induced dyskinesia [9]. While this information is insightful, current studies are inconclusive with no actionable guidance for medication adjustment. More studies in the area of how DRD polymorphisms translate to dosing strategies are needed. Therefore, clinical decisions should still prioritize symptom management and adverse effect mitigation through standard therapeutic adjustments.

Summary of Recommendations

- Consider increasing the dose of rasagiline to 1 mg daily to help augment the dopaminergic effect and reduce the "off symptoms" the patient is experiencing.
- Consider adding an ER formulation of carbidopa/levodopa that may improve bioavailability by up to 70%. Careful consideration to avoid increasing the L-dopa dose if the patient has experienced peak dose dyskinesias in the past.
- Review patient's medication fill history and adherence to all his medications including opicapone (COMT inhibitor), which usually helps to extend the dopaminergic effect of dopamine.
- Dopaminergic medications will increase the risk of orthostatic hypotension. Consider replacing medications with the highest anticholinergic burden and effect on GI absorption that may impact motor complications and worsen hypotension. An example would be switching paroxetine to escitalopram (he is *CYP2C19* normal metabolizer).

Discussion with the Patient and Family

It is important to communicate the limitations of PGx testing in this case. Various PGx studies have been conducted looking at different genes associated with drug response and side effects, but the up-to-date data have been inconclusive. The focus should then be on carefully adjusting medications based on their symptoms and tolerability, with regular follow-ups to assess improvements or adverse effects. Managing expectations and ensuring the family is aware of the balance between symptom control and potential adverse effects, such as anticholinergic burden and orthostatic hypotension, is key to shared decision-making.

What Did PGx Change About the Case Outcome?

- While there is still ongoing research developing around the PGx associated with levodopa-related response and side effects, clinical application remains limited and requires further research.
- Consider well-established PGx markers such as *CYP2D6* for guiding therapy evaluation as it can impact on drug metabolism, affecting medications such as paroxetine, which can have downstream side effects.

References

1. 2023 American Geriatrics Society Beers Criteria® Update Expert Panel. American Geriatrics Society 2023 updated AGS Beers Criteria® for potentially inappropriate medication use in older adults. J Am Geriatr Soc. 2023;71(7):2052–81. https://doi.org/10.1111/jgs.18372. Epub 2023 May 4. PMID: 37139824
2. Achterberg W, Lautenbacher S, Husebo B, Erdal A, Herr K. Pain in dementia. Pain Rep. 2020;5(1):e803. https://doi.org/10.1097/pr9.0000000000000803.
3. Administration US FDA. NORTHERA® (droxidopa) capsules, for oral use. Prescribing Information; 2017.
4. American Geriatrics Society. updated AGS Beers Criteria® for potentially inappropriate medication use in older adults (2023). J Am Geriatr Soc. 2023;71(7):2052–81. https://doi.org/10.1111/jgs.18372.
5. Bishop SM Jr, Butler T, Carpentier RM, Heins KR, Formea CM. Pharmacogenomics to support mental health medication therapy management: clinical practice considerations and a conceptual framework to enhance patient care. J Am Coll Clin Pharm. 2024;7(2) https://doi.org/10.1002/jac5.1892.
6. Bousman CA, Stevenson JM, Ramsey LB, Sangkuhl K, Hicks JK, Strawn JR, Singh AB, Ruaño G, Mueller DJ, Tsermpini EE, Brown JT, Bell GC, Leeder JS, Gaedigk A, Scott SA, Klein TE, Caudle KE, Bishop JR. Clinical Pharmacogenetics implementation consortium (CPIC) guideline for CYP2D6, CYP2C19, CYP2B6, SLC6A4, and HTR2A genotypes and serotonin reuptake inhibitor antidepressants. Clin Pharmacol Ther. 2023;114(1):51–68. https://doi.org/10.1002/cpt.2903.
7. Brucki SMD, Aprahamian I, Borelli WV, da Silveira VC, Ferretti CEL, Smid J, Barbosa B, Schilling LP, Balthazar MLF, Frota NAF, de Souza LC, Vale FAC, Caramelli P, Bertolucci PHF, Chaves MLF, Nitrini R, Schultz RR, Morillo LS. Management in severe dementia: recommendations of the scientific Department of Cognitive Neurology and Aging of the Brazilian academy of neurology. Dement Neuropsychol. 2022;16(3 Suppl 1):101–20. https://doi.org/10.1590/1980-5764-dn-2022-s107pt.
8. Cacabelos R. Pharmacogenetic considerations when prescribing cholinesterase inhibitors for the treatment of Alzheimer's disease. Expert Opin Drug Metab Toxicol. 2020;16(8):673–701. https://doi.org/10.1080/17425255.2020.1779700.
9. Chaparro-Solano HM, Rivera Paz M, Anis S, Hockings JK, Kundrick A, Piccinin CC, Assaedi E, Saadatpour L, Mata IF. Critical evaluation of the current landscape of pharmacogenomics in Parkinson's disease – what is missing? A systematic review Parkinsonism. Relat Disord. 2025;130:107206. https://doi.org/10.1016/j.parkreldis.2024.107206.
10. Costa B, Vale N. Understanding lamotrigine's role in the CNS and possible future evolution. Int J Mol Sci. 2023;24(7) https://doi.org/10.3390/ijms24076050.

11. Deprescribing Guide for Anticholinergic Drugs for Urinary Incontinence (Antimuscarinics). NSW Government. 2019. https://www.nswtag.org.au/wp-content/uploads/2018/06/1.6-Deprescribing-Guide-for-Anticholinergic-drugs-for-Urinary-Incontinence-Antimuscarinics.pdf. Accessed Jan 19 2025.
12. Desai AK, Grossberg GT. Recognition and management of behavioral disturbances in dementia. Prim Care Companion J Clin Psychiatry. 2001;3(3):93–109. https://doi.org/10.4088/pcc.v03n0301.
13. Dumurgier J, Tzourio C. Epidemiology of neurological diseases in older adults. Rev Neurol (Paris). 2020;176(9):642–8. https://doi.org/10.1016/j.neurol.2020.01.356.
14. Feigin VL, Vos T, Nichols E, Owolabi MO, Carroll WM, Dichgans M, Deuschl G, Parmar P, Brainin M, Murray C. The global burden of neurological disorders: translating evidence into policy. Lancet Neurol. 2020;19(3):255–65. https://doi.org/10.1016/s1474-4422(19)30411-9.
15. Ferreira JJ, Lees A, Rocha JF, Poewe W, Rascol O, Soares-da Silva P. Long-term efficacy of opicapone in fluctuating Parkinson's disease patients: a pooled analysis of data from two phase 3 clinical trials and their open-label extensions. Eur J Neurol. 2019;26(7):953–60. https://doi.org/10.1111/ene.13914. Epub 2019 Mar 25. PMID: 30681754; PMCID: PMC6593852
16. Guerreiro R, Hardy J. Genetics of Alzheimer's disease. Neurotherapeutics. 2014;11(4):732–7. https://doi.org/10.1007/s13311-014-0295-9.
17. Karnes JH, Rettie AE, Somogyi AA, et al. Clinical Pharmacogenetics Implementation Consortium (CPIC) Guideline for CYP2C9 and HLA-B Genotypes and Phenytoin Dosing: 2020 Update. Clin Pharmacol Ther. 2021;109(2):302–9. https://doi.org/10.1002/cpt.2008. Epub 2020 Sep 6. PMID: 32779747; PMCID: PMC7831382
18. Kuz GM, Manssourian A. Carbamazepine-induced hyponatremia: assessment of risk factors. Ann Pharmacother. 2005;39(11):1943–6. https://doi.org/10.1345/aph.1G209.
19. Manson LEN, Nijenhuis M, Soree B, de Boer-Veger NJ, Buunk AM, Houwink EJF, Risselada A, Rongen G, van Schaik RHN, Swen JJ, Touw DJ, van Westrhenen R, Deneer VHM, Guchelaar HJ. Dutch Pharmacogenetics Working Group (DPWG) guideline for the gene-drug interaction of CYP2C9, HLA-A and HLA-B with anti-epileptic drugs. Eur J Hum Genet. 2024;32(8):903–11. https://doi.org/10.1038/s41431-024-01572-4.
20. Mirabegron package insert. Ahmedabad, India: Zydus Lifesciences, Ltd; 2024. Accessed July 21, 2025.
21. McDermott CL, Gruenewald DA. Pharmacologic management of agitation in patients with dementia. Curr Geriatr Rep. 2019;8(1):1–11. https://doi.org/10.1007/s13670-019-0269-1.
22. Narayan P, Tunuguntla HS. Long-term efficacy and safety of tamsulosin for benign prostatic hyperplasia. Rev Urol. 2005;7 Suppl 4(Suppl 4):S42–8.
23. Phillips EJ, Sukasem C, Whirl-Carrillo M, Müller DJ, et al. Clinical Pharmacogenetics Implementation Consortium Guideline for HLA genotype and use of carbamazepine and oxcarbazepine: 2017 update. Clin Pharmacol Ther. 2018;103(4):574–81. https://doi.org/10.1002/cpt.1004. Epub 2018 Feb 2. PMID: 29392710; PMCID: PMC5847474
24. Rascol O, et al. Rasagiline as an adjunct to levodopa in patients with Parkinson's disease and motor fluctuations (LARGO, lasting effect in adjunct therapy with Rasagiline given once daily, study): a randomised, double-blind, parallel-group trial. Lancet. 2005;365(9463):947–54.
25. Tamsulosin (package insert). Bridgewater, NJ: Amneal Pharmaceuticals, LLC; 2025. Accessed July 21, 2025)
26. Theken KN, Lee CR, Gong L, Caudle KE, Formea CM, Gaedigk A, Klein TE, Agúndez JAG, Grosser T. Clinical Pharmacogenetics implementation consortium guideline (CPIC) for CYP2C9 and nonsteroidal anti-inflammatory drugs. Clin Pharmacol Ther. 2020;108(2):191–200. https://doi.org/10.1002/cpt.1830.
27. Wancata J, Alexandrowicz R, Marquart B, et al. The criterion validity of the geriatric depression scale: a systematic review. Acta Psychiatr Scand. 2006;114:398–410.

Chapter 8
Pharmacogenomics and Psychiatry Care for Older Adults

Clayton English, David Bright, and Yee Ming Lee

Introduction

Older adults are vulnerable to experiencing many similar mental health challenges that younger populations develop; however, physiological changes related to aging can have a major influence on clinical management. Outside of neurocognitive disorders, depression and anxiety are two of the most common psychiatric disorders encountered in older adults [1]. Major depression is highly prevalent in both long-term care settings and general community settings but is less common compared to younger populations (2% versus 6.6%) [2–4]. Late-life depression can occur among individuals 65 years and older with no previous history of depression [5]. However, many older patients experience depressive symptoms that are clinically meaningful while not fulfilling all of the required diagnostic criteria for major depression [4]. Untreated major depression and depressive symptoms in older adults can lead to poor outcomes including increased disability, decline in functional status, and a higher mortality rate. Despite this, depression in older adults is often underdiagnosed and undertreated [6–8]. Lack of formal diagnosis can be secondary to multiple factors, including misconceptions about the normal aging process, ageist practices in healthcare, stigma associated with mental illness, and physical

C. English (✉)
School of Pharmacy, University of Washington, Seattle, WA, USA
e-mail: claytone@uw.edu

D. Bright
Professor of Pharmaceutical Sciences, College of Pharmacy, Ferris State University, Big Rapids, MI, USA

Y. M. Lee
Associate Professor of Clinical Pharmacy, Skaggs School of Pharmacy and Pharmaceutical Sciences, University of Colorado, Aurora, CO, USA

T. DeLellis et al. (eds.), *Applying Pharmacogenomics to Optimize Geriatric Care*, https://doi.org/10.1007/978-3-032-10653-7_8

manifestations of depression being misdiagnosed [9]. Furthermore, suicide is a significant risk factor for older adults, especially those experiencing depression with psychotic features or new-onset disability [10–12].

Although biological age-related changes in the brain such as neurodegeneration and decline in select monoaminergic transmission (e.g., reduction in norepinephrine) signal that depression may increase with age, chronological age (i.e., the number of years lived) is not, in itself, a well-defined risk factor for late-life depression [9, 13–15]. Having a comorbid chronic medical illness is a strong risk factor for developing depression, which older adults are more likely to have, therefore increasing susceptibility [16]. The diagnostic criteria for depression are not different in older adults; however, depressive subtypes and symptomology tend to be different in older adults relative to younger cohorts. Psychotic symptoms are more frequently encountered and diagnosed in older adults [12, 17]. Anhedonia (i.e., loss of interest in pleasurable activities) and somatization (i.e., hyperfocus on physical complaints) both tend to be more prevalent in older adults and may present without profound changes in mood or sadness [9, 17]. Depressive symptoms may also originate from other biological sequelae including vascular lesion development or collaterally from dementia onset and progression [18]. Clinical trial design and outcomes for depression are also different for older adults. Effect sizes are often comparable to those observed in younger populations; however, antidepressant effects may weaken as age increases [4]. The duration of treatment is often longer in older adults, typically 12 weeks compared to 6 weeks [4, 9]. Additionally, many older patients are excluded from studies including the oldest patients (≥ 85 years of age), those with suicidality, and medically fragile patients, which may limit the generalizability of antidepressant clinical trials conducted in older adults. Conservative approaches in treatment may lead to underdosing of medications or contribute to medications being withheld [4]. Additional complexity arises in optimizing treatment because of medical comorbidities and drug–drug interactions that are more commonly seen in older populations. Unfortunately, psychotropic medications, including antidepressants, are one of the leading causes of preventable adverse sequelae in older adults; thus, vigilant oversight in the clinical utility of these medications is a priority [19].

Given the challenges facing clinicians in treating depression in older adults, pharmacogenomics may provide additional information to serve as a decision-support tool for pharmacotherapy selection in combination with other patient factors, including medical comorbidities, drug–drug interactions, and age-related physiologic changes. Routine pharmacogenomic testing is currently not endorsed in treatment guidelines for depression due to the limited efficacy data. However, it can provide support to clinicians in optimizing dosing, guiding medication selection, and informing next-best steps in tailoring therapy for those who failed initial treatment, have refractory forms of illness, or experience treatment-emergent adverse effects [20–22]. The International Society of Psychiatric Genetics and the Clinical Pharmacogenetics Implementation Consortium (CPIC) Guidelines provide detailed recommendations regarding select antidepressants that have phenotypic variability in metabolism and clearance, with some phenotypes, such as poor, intermediate, or ultrarapid metabolizers, impacting clinical outcomes [20, 23, 24]. Select antidepressants addressed in the guidelines are predominantly metabolized by CYP2B6, CYP2D6, and CYP2C19 pathways. Although guidelines do not address all antidepressants, knowledge of CYP450 phenotypes can provide useful information in tailoring treatment, especially in older adults who may already have reduced oxidative metabolism secondary to aging.

Pharmacogenomics of Antidepressants: The Roles of CYP2D6, CYP2C19, and CYP2B6

Genetic variations in *CYP2D6, CYP2C19,* and *CYP2B6* could influence the metabolism of many antidepressants and may potentially affect dosing, efficacy, and tolerability of these drugs.

Citalopram, Escitalopram, Sertraline CYP2C19 metabolizes citalopram and escitalopram to less potent metabolites, with CPIC recommendations shown in Table 8.1a [23]. The FDA label limits the dose of citalopram to 20 mg per day in CYP2C19 poor metabolizers due to the risk of QT prolongation; as well as in patients older than 60 years of age; patients with hepatic impairment, or those taking CYP2C19 inhibitors (e.g., fluconazole, fluoxetine) [25]. While the FDA label of escitalopram does not include a dose limit for CYP2C19 poor metabolizers, QT prolongation can occur and may warrant monitoring for those with risk factors. Sertraline is metabolized by several pathways including CYP2C19 (major pathway) and CYP2B6. CPIC guideline provides dose recommendations for sertraline when CYP2C19 result is available (Table 8.1b), and when both CYP2C19 and CYP2B6 results are available (Table 8.1c).

Fluvoxamine, Paroxetine, Vortioxetine, Venlafaxine CYP2D6 metabolizes fluvoxamine, paroxetine, and vortioxetine into less active metabolites, while it metabolizes venlafaxine (active parent drug) to an active metabolite, desvenlafaxine. CPIC guideline dose recommendations for these drugs based on CYP2D6 phenotypes are shown in Table 8.1d.

Tricyclic Antidepressants (TCA) TCAs can be classified by their chemical structure into tertiary amines (such as amitriptyline, clomipramine, doxepin, imipramine, and trimipramine), and secondary amines (such as desipramine and nortriptyline). Tertiary amines are mainly metabolized by CYP2C19 to secondary amines. Amitriptyline and imipramine are metabolized by CYP2C19 to pharmacologically active metabolites of nortriptyline and desipramine, respectively. Both tertiary and secondary amines are metabolized by CYP2D6 to less active metabolites. CPIC guidelines provide dosing recommendations for tertiary amines based on the different combination of CYP2C19 and CYP2D6 phenotypes (Table 8.2a) and for secondary amines and CYP2D6 phenotypes (Table 8.2b). Of note, CPIC guideline recommendations for TCAs apply to higher TCA doses to treat conditions such as

Table 8.1 CPIC guideline for CYP2D6, CYP2C19, CYP2B6, SLC6A4, and HTR2A genotypes and serotonin reuptake inhibitor antidepressants [23]

(a) Citalopram, escitalopram, and CYP2C19

CYP2C19 phenotype	CPIC recommendations
Ultrarapid metabolizer	Consider using an alternative antidepressant not predominantly metabolized by CYP2C19. If citalopram or escitalopram use are warranted, consider using a higher maintenance dose if adequate efficacy is not achieved at standard maintenance dose.
Rapid metabolizer	Initiate at the recommended starting dose. If patient does not respond adequately, consider using a higher maintenance dose or use an alternative antidepressant not predominantly metabolized by CYP2C19.
Normal metabolizer	Initiate at the recommended starting dose.
Intermediate metabolizer	Initiate at the recommended starting dose, and titrate slowly and a lower maintenance dose.
Poor metabolizer	Consider using an alternative antidepressant not predominantly metabolized by CYP2C19. If citalopram or escitalopram use are warranted, consider a lower starting dose, titrate slowly, and a 50% reduction in standard maintenance dose. Per FDA label, the maximum recommended dose of citalopram is 20 mg per day in CYP2C19 poor metabolizers due to the risk of QT prolongation. The label also limits citalopram to 20 mg per day dose in patients >60 years of age, those with liver impairment, and those taking CYP2C19 inhibitors.

(b) Sertraline and CYP2C19

CYP2C19 phenotype	CPIC recommendations
Ultrapid metabolizer	Initiate at the recommended starting dose
Rapid metabolizer	
Normal metabolizer	
Intermediate metabolizer	Initiate at the recommended starting dose. Consider slower titration, and lower maintenance dose.
Poor metabolizer	Consider a lower starting dose, slower titration, and 50% reduction in maintenance dose, or use an alternative drug not predominantly metabolized by CYP2C19

(c) Sertraline and CYP2B6

CYP2B6 Phenotype	CPIC recommendations
Ultrarapid metabolizer	Initiate at the recommended starting dose
Rapid metabolizer	
Normal metabolizer	

(continued)

Table 8.1 (continued)

(c) Sertraline and CYP2B6	
CYP2B6 Phenotype	CPIC recommendations
Intermediate metabolizer	Initiate at the recommended starting dose. Consider slower titration, and lower maintenance dose.
Poor metabolizer	Consider a lower starting dose, slower titration, and 25% reduction in maintenance dose, or use an alternative drug not predominantly metabolized by CYP2B6

(d) Sertraline and CYP2B6 and CYP2C19

Phenotype	CYP2B6 ultrarapid or rapid metabolizer	CYP2B6 normal metabolizer	CYP2B6 intermediate metabolizer	CYP2B6 poor metabolizer
CYP2C19 ultrarapid or rapid metabolizer	Initiate at the recommended starting dose. If patient does not respond adequately, consider using a higher maintenance dose or use an alternative antidepressant not predominantly metabolized by CYP2C19 or CYP2B6.	Initiate at the recommended starting dose	Initiate at the recommended starting dose	Initiate at the recommended starting dose
CYP2C19 normal metabolizer	Initiate at the recommended starting dose	Initiate at the recommended starting dose	Initiate at the recommended starting dose, consider slow titration, and a lower maintenance dose	Consider a lower starting dose, titrate slowly with a 25% reduction in maintenance dose or use an alternative drug not predominantly metabolized by CYP2B6
CYP2C19 intermediate metabolizer	Initiate at the recommended starting dose	Initiate at the recommended starting dose, consider slow titration, and a lower maintenance dose	Initiate at the recommended starting dose, consider slow titration and a lower maintenance dose	Consider a lower starting dose, titrate slowly with a 50% reduction in maintenance dose or use an alternative drug not predominantly metabolized by CYP2B6

(continued)

Table 8.1 (continued)

(d) Sertraline and CYP2B6 and CYP2C19

Phenotype	CYP2B6 ultrarapid or rapid metabolizer	CYP2B6 normal metabolizer	CYP2B6 intermediate metabolizer	CYP2B6 poor metabolizer
CYP2C19 poor metabolizer	Consider a lower starting dose, titrate slowly with a 50% reduction in maintenance dose or use an alternative drug not predominantly metabolized by CYP2B6	Consider a lower starting dose, titrate slowly with a 50% reduction in maintenance dose or use an alternative drug not predominantly metabolized by CYP2B6	Consider a lower starting dose, titrate slowly with a 50% reduction in maintenance dose or use an alternative drug not predominantly metabolized by CYP2B6	Use an alternative antidepressant not predominantly metabolized by CYP2C19 or CYP2B6

(e) Fluvoxamine, paroxetine, venlafaxine, vortioxetine, and CYP2D6

CYP2D6 phenotype (activity score; AS)	CPIC recommendations
Ultrarapid metabolizer (AS >2.25)	Fluvoxamine: Initiate at the recommended starting dose[a] Paroxetine: Select an alternative drug not predominantly metabolized by CYP2D6 Venlafaxine: Initiate at the recommended starting dose[a] Vortioxetine: Use an alternative drug not predominantly metabolized by CYP2D6. If vortioxetine is warranted, initiate at recommended starting dose and titrate based on response and side effects
Normal metabolizer (AS 1.25–2.25)	Fluvoxamine: Initiate at the recommended starting dose Paroxetine: Initiate at the recommended starting dose Venlafaxine: Initiate at the recommended starting dose Vortioxetine: Initiate at the recommended starting dose
Intermediate metabolizer (AS 0.25–1)	Fluvoxamine: Initiate at the recommended starting dose Paroxetine: Consider a lower starting dose and titrate slowly Venlafaxine: Initiate at the recommended starting dose[a] Vortioxetine: Initiate at the recommended starting dose
Poor metabolizer (AS 0)	Fluvoxamine: Consider a 25–50% lower starting dose and titrate slowly; or use an alternative drug not predominantly metabolized by CYP2D6 Paroxetine: Consider a 50% reduction in the recommended starting dose, titrate slowly to a 50% lower maintenance dose Venlafaxine: Consider using an alternative antidepressant not predominantly metabolized by CYP2D6 Vortioxetine: Initiate 50% of the starting dose (e.g., 5 mg) and titrate to a maximum recommended dose of 10 mg per day or use an alternative antidepressant not predominantly metabolized by CYP2D6

[a]CPIC has no specific guidance due to limited data

depression. As TCAs are also prescribed at lower doses for conditions such as neuropathic pain, CPIC guideline does not provide dose recommendations for these uses as patients with altered CYP2C19 or CYP2D6 phenotypes are less likely to experience side effects at lower doses of TCA.

Table 8.2 CPIC guideline for CYP2D6 and CYP2C19 genotypes and dosing of tricyclic antidepressants: 2016 update [55]

(a) CPIC guideline recommendations for amitriptyline based on CYP2D6 and CYP2C19 phenotypes. This can be applied to other tertiary amines, including clomipramine, doxepin, imipramine, and trimipramine

Phenotype	CYP2D6 ultrarapid metabolizer	CYP2D6 normal metabolizer	CYP2D6 intermediate metabolizer	CYP2D6 poor metabolizer
CYP2C19 ultrarapid or rapid metabolizer	Avoid amitriptyline use	Consider using an alternative antidepressant not metabolized by CYP2C19	Consider using an alternative antidepressant not metabolized by CYP2C19	Avoid amitriptyline use
CYP2C19 normal metabolizer	Avoid amitriptyline use. If amitriptyline is warranted, consider titrating to a higher dose.	Initiate at the recommended starting dose	Consider a 25% reduction in the recommended starting dose	Avoid amitriptyline use. If amitriptyline use is warranted, consider a 50% reduction in recommended starting dose.
CYP2C19 intermediate metabolizer	Avoid amitriptyline use	Initiate at the recommended starting dose	Consider a 25% reduction in the recommended starting dose.	Avoid amitriptyline use. If amitriptyline use is warranted, consider a 50% reduction in the recommended starting dose.
CYP2C19 poor metabolizer	Avoid amitriptyline use	Avoid amitriptyline use. If amitriptyline use is warranted, consider a 50% reduction in the recommended starting dose.	Avoid amitriptyline use	Avoid amitriptyline use

(b) CPIC recommendations for secondary amine tricyclic antidepressants such as nortriptyline and desipramine based on CYP2D6 phenotype

CYP2D6 phenotype (activity score; AS)	CPIC recommendations
Ultrarapid metabolizer (AS >2.25)	Avoid tricyclic use due to potential lack of efficacy. Use an alternative antidepressant not predominantly metabolized by CYP2D6. If a TCA is warranted, consider titrating to a higher target dose or utilize therapeutic drug monitoring to guide dose adjustment.
Normal metabolizer (AS 1.25–2.25)	Initiate at the recommended starting dose

(continued)

Table 8.2 (continued)

(b) CPIC recommendations for secondary amine tricyclic antidepressants such as nortriptyline and desipramine based on CYP2D6 phenotype	
CYP2D6 phenotype (activity score; AS)	CPIC recommendations
Intermediate metabolizer (AS 0.25–1)	Consider a 25% dose reduction in the recommended starting dose. Utilize therapeutic drug monitoring to guide dose adjustments.
Poor metabolizer (AS 0)	Avoid tricyclic use due to potential for side effects. Use an alternative antidepressant not predominantly metabolized CYP2D6. If a TCA is warranted, consider a 50% reduction of recommended starting dose. Utilize therapeutic drug monitoring to guide dose adjustment.

Other Antidepressants Fluoxetine is metabolized by CYP2D6 to an active metabolite, S-norfluoxetine with similar activity. Although CYP2D6 genetic polymorphisms alter the fluoxetine and norfluoxetine parent-to-metabolite ratio, the sum total fluoxetine and norfluoxetine plasma levels did not vary significantly based on CYP2D6 metabolizer phenotype; hence, there is no CPIC recommendation available for CYP2D6 and fluoxetine. Duloxetine, on the other hand, is metabolized by CYP1A2 and CYP2D6, but current data is limited to derive any pharmacogenomic recommendations.

CYP2D6 Phenoconversion Caused by Certain Antidepressants

Aside from reviewing patient's genotype results to guide antidepressant selection, one must consider the potential for phenoconversion caused by the new antidepressant, which can affect the metabolism of patient's existing medications. As discussed in Chap. 2, phenoconversion occurs when the clinical phenotype does not match the genotype-based phenotype [26]. In particular, CYP2D6 phenoconversion is prevalent with the use of psychotropics, with 20–30% of patients taking a CYP2D6 inhibitor that can cause phenoconversion. This includes strong CYP2D6 inhibitors (i.e., bupropion, fluoxetine, and paroxetine) and moderate CYP2D6 inhibitors (i.e., duloxetine, mirabegron). For instance, a patient who is a CYP2D6 ultrarapid, normal or intermediate metabolizer is initiated on fluoxetine (a strong CYP2D6 inhibitor), patient phenoconverts to behave as a CYP2D6 poor metabolizer of the existing medications that are CYP2D6-dependent. The potential for CYP2D6 phenoconversion caused by the new antidepressant does not preclude its use. One may have to adjust the dose of existing CYP2D6-dependent medications to reduce the risk of side effects caused by increased drug levels. On the other hand, CYP2D6 phenoconversion does not occur in patients who are already CYP2D6 poor metabolizers based on their genotype. As antidepressant switching can occur, the removal of CYP2D6 inhibitor results in patient reverting back to their genotype-based phenotype. Providers may have to readjust patient's CYP2D6-dependent medications with the change in antidepressants. As phenoconversion is reversible, it usually takes 5 to 7 days after discontinuing the inhibitor to revert back to patient's genotype-based phenotype [27].

The following three patient cases represent common scenarios encountered in the management of depression and other psychiatric disorders in older adults. These cases can help practitioners understand the role, clinical utility, and limitations of pharmacogenomic testing in optimizing pharmacotherapy management for psychiatric disorders in older adults.

Case 1

A 77-year-old female presents to her primary care physician's office for a follow-up visit six months after a fall that resulted in a left ulnar fracture. The fracture seems to have healed well, and the patient is no longer in pain associated with the arm, although she complains that her injury slowed her down during the healing process. She feels like she has less energy, less motivation, trouble falling asleep/maintaining sleep duration, and is more socially isolated than before the injury given that she couldn't participate in the same activities that she did before the injury.

Medical History:

Depression (22 years)
Hypertension (31 years)
Osteoporosis (12 years)
Gastroesophageal reflux disease (11 years)

Allergies:

None known

Pertinent Vitals and Labs:

BP: 120/80, HR: 60, Temp: 37 °C, RR:14

	9 months ago	Today
Patient health questionnaire (PHQ)-9	6 (mild)	10 (moderate)

Pharmacogenomics Results:

*CYP2D6 *1/*1* normal metabolizer
*CYP2C9 (*1/*3, AS 1)* intermediate metabolizer
*CYP2C19 *17/*17* ultrarapid metabolizer
*CYP2B6 *1/*1* normal metabolizer

Current Medications:

Hydrochlorothiazide 25 mg daily
Alendronate 70 mg weekly

Acetaminophen 1000 mg twice daily as needed for arm/wrist pain
Citalopram 10 mg daily
Famotidine 20 mg twice daily

Clinical (Nongenetic) Considerations

The treatment goals for this patient are to improve target symptoms of depression by reducing both anhedonia and avolition and improving her sleep cycle and fatigue. The injury, associated recovery, and its impact on daily life may have resulted in a decline in mental health functioning, particularly in the context of a prior diagnosis of depression.

Treatment guidelines for depression are limited in their guidance in tailoring therapy for older adults. There are multiple treatments that have evidence in treating late-life depression including selective serotonin reuptake inhibitors (SSRIs) serotonin-norepinephrine reuptake inhibitors (SNRIs), TCAs, mirtazapine, and non-pharmacological treatment options [28–30]. Non-pharmacological options should be considered for many patients with depression, including evidence-based psychotherapy such as cognitive behavioral therapy, exercise, and light therapy [28]. Depression is fundamentally challenging to manage due to the high treatment failure rate at baseline, the need for antidepressant dose titration, and considerations regarding drug–drug–interactions. Furthermore, the slow onset of therapeutic effect prolongs the time to symptom remission and is extended further by physiologic aging.

The most common pharmacotherapeutic options are the SSRIs and SNRIs, which are commonly utilized given their wide-safety index, ability to target comorbid conditions such as anxiety and clinical benefits of treating depression at starting doses. While SSRIs and SNRIs are relatively better tolerated compared to older treatments, including TCAs and monoamine oxidase inhibitors (MAOIs), treatment-emergent side effects remain common in older adults, who are often more susceptible to pharmacodynamic adverse effects such as hyponatremia and bleeding [28]. Mirtazapine and bupropion are also considered first-line therapies for depression; however, the evidence for their use in older adults is weaker compared to SSRIs since there is a lack of placebo-controlled trials demonstrating their efficacy in this population [29]. Positive comparator-controlled trial data to paroxetine exists for both [31, 32]. Mirtazapine and bupropion are predominantly used if there are compelling symptoms that can be matched to their corresponding side effect profile to improve select target symptoms of depression and overall tolerability. For example, individuals struggling with insomnia or poor appetite could benefit from mirtazapine's side effect profile, which includes appetite stimulation, weight gain, and sedation. Likewise, older adults with anhedonia, fatigue, and concentration difficulties might benefit from bupropion's activating properties.

Many antidepressants are listed on the AGS Beers Criteria® as being inappropriate for older adults. These include TCAs, which are generally avoided because of their anticholinergic, histaminergic, and alpha-1 antagonistic effects. TCAs are associated with falls and cognitive impairment. Paroxetine is also not recommended because of its anticholinergic activity [25]. Although TCAs should not be used in most older adults given their safety profile, there are some caveats to this rule. It should be noted that nortriptyline has been well studied in late-life depression and can be considered for refractory cases of late-life depression [29, 33]. TCAs can also be used for underlying pain conditions and can be considered for select cases of depression that present comorbid with pain. Therapeutic drug monitoring, if available, can assist in optimizing plasma concentrations for patients, especially for those with altered CYP2D6 metabolism. Paroxetine has also been formally studied in late-life depression but given the variety of safer SSRIs with lower anticholinergic burden, it is generally only used if there is a compelling reason, such as previous history of positive response.

The therapeutic options for this patient include increasing the citalopram dose, switching to a different SSRI (e.g., sertraline), SNRI (e.g., duloxetine, venlafaxine), mirtazapine, bupropion, vortioxetine, or a TCA. Among the SSRIs, sertraline can be considered; fluoxetine may improve patient's fatigue, while paroxetine is less preferred due to its anticholinergic effects in older adults. Among the SNRIs, duloxetine and venlafaxine are options if her blood pressure is well-controlled. If a treatment is used that blocks the serotonin transporter, closer monitoring of sodium values should be implemented because of hyponatremia risks. Bupropion may improve the patient's fatigue, while mirtazapine may aid her sleep. TCAs are less preferred given their side effect profile in older adults.

Pharmacogenomic Considerations

Citalopram is metabolized primarily by CYP450 3A4 and 2C19. This patient has a CYP2C19 ultrarapid metabolizer phenotype, resulting in lower citalopram plasma concentrations than the normal CYP2C19 metabolizer phenotypes. This increases the risk of treatment failure given the lower therapeutic levels. One option would be to increase the citalopram dose until symptoms improve if it remains as the preferred treatment agent. However, it may take a higher than typical dose to achieve symptom resolution in this patient given the pharmacogenomic data. Clinicians should be mindful of citalopram's ability to prolong the QT interval secondary to its didemethylcitalopram (DDCT) metabolite produced via CYP2D6 metabolism of citalopram's initial metabolite, demethylcitalopram [34]. Given the patient has a CYP2D6 normal metabolizer phenotype, DDCT concentrations would theoretically be expected to be normal, and the patient would still be susceptible to dose-related increases in QT interval prolongation despite the CYP2C19 ultrarapid metabolizer phenotype. FDA-labeling recommends individuals over 60 years of age should not receive doses of citalopram greater than 20 mg/day, which may serve as a barrier to uptitrating the citalopram dose based on the patient's CYP2C19 results [35].

Alternatively, another SSRI agent that is not predominantly metabolized by CYP2C19 to consider is sertraline as it is metabolized by both CYP2C19 and CYP2B6. CPIC guideline recommendation (see Table 8.1d) for sertraline in this patient, who has CYP2C19 ultrarapid and CYP2B6 normal metabolizer phenotypes, is to initiate sertraline at the recommended starting dose per FDA labeling. Her CYP2D6 normal metabolizer phenotype would suggest that an alternative antidepressant (e.g., fluoxetine, venlafaxine, vortioxetine) would be expected to exhibit a typical dose/concentration relationship when usual doses are prescribed. If duloxetine is considered, it is metabolized via CYP1A2 and CYP2D6. There is currently no pharmacogenomic guideline available for duloxetine, so the usual dose can be initiated.

Regardless of the metabolism status, some pharmacodynamic side effects are not well predicted, and the risk is still present. Hyponatremia secondary to syndrome of inappropriate antidiuretic hormone (SIADH) release can occur with all serotonin transport blocking agents when used in older populations and should be monitored regardless of the pharmacogenomic findings.

Overall Recommendations

- If a dose increase in citalopram is not preferred by the patient, symptomatic reductions are not achieved, or QT prolongation is observed by increasing the dose, switching therapies should be considered.
- Given the pharmacogenomic profile and lack of significant compelling indications or depressive subtypes, several other treatments could be considered including sertraline, an SNRI, mirtazapine, or bupropion.
- Using shared decision making and discussing the potential benefits as well as drawbacks of each treatment should be prioritized with the patient. Given that significant efficacy differences between classes for late-life depression have not been firmly established, both the patient and clinician have considerable leeway in treatment selection and management.

What Did Pharmacogenomics Change About the Case Outcome?

- CYP2C19 ultrarapid metabolism of the patient's citalopram may have contributed to poor dose optimization and ineffective treatment outcomes for citalopram.
- Identifying the patient's CYP2C19 phenotypes through pharmacogenomic testing provided a reason for optimizing the citalopram dose, but increasing the dose beyond the typical dosing target for older adults could be a limitation based on her CYP2D6 phenotype unless the risk for QT prolongation is appropriately monitored and addressed.

- Pharmacogenomic testing further provided clear guidance for selection and dosing of alternate medication options, reducing the risk that a similar pharmacogenomic issue could arise with an SSRI also predominantly metabolized by CYP2C19, such as escitalopram.
- Pharmacogenomic testing did not alter treatment and dosing guidance for the majority of alternative first-line antidepressants used in older adults.

Case 2

A 70-year-old male presents to his primary care for his annual review. His primary care provider is concerned about the patient's chronic use of clonazepam for his anxiety and insomnia and wants to taper down his benzodiazepine. The patient is very hesitant to consider this option due to concerns of worsening anxiety and insomnia symptoms, which he has struggled with for decades. The patient's anxiety and insomnia symptoms continue to remain uncontrolled, predominantly complicated by having difficulty falling asleep, muscle tension, and excessive worries. The patient just completed pharmacogenomic testing with results available. The primary care physician orders a pharmacy consult for assistance with anxiety management.

Medical History:

Generalized anxiety disorder (GAD)
Hypertension
Hypothyroidism
Heart failure
Chronic kidney disease Stage III
Type 2 diabetes
Insomnia
Tension headache
Mild cognitive impairment
Osteoarthritis of the knee

Allergies:

None known

Pertinent Vitals and Labs:

140 lb., 5 ft. 5 in, BMI 23.3, BP 140/80, HR 75, eGFR 50 ml/min, K+ 4 mM, EF 40%, TSH 4.32
EKG: normal sinus rhythm

	3 months ago	Today
Generalized anxiety disorder (GAD)-7 score	13 (moderate)	10 (moderate)
PHQ-9 score	6 (mild)	4 (normal)

Pharmacogenomic Results:

*CYP2C19 *1/*1* normal metabolizer
*CYP2D6 *1/*4* intermediate metabolizer
*CYP2B6 *1/*6* intermediate metabolizer

Current Medications:

Acetaminophen 325 mg one to two tablets as needed for headaches or osteoarthritis pain
Clonazepam 0.5–1 mg twice daily as needed for anxiety and sleep
Empagliflozin 10 mg once daily
Furosemide 40 mg twice daily
Hydroxyzine 25–50 mg three times daily as needed for anxiety
Levothyroxine 75 mcg daily
Lisinopril 10 mg once daily
Metoprolol succinate XL 100 mg once daily
Topical diclofenac 1% gel: Apply a small amount four times a day as needed for osteoarthritis pain of the knee

Clinical (Nongenetic) Considerations

The treatment goals for this patient are to improve target symptoms of anxiety and insomnia, while simultaneously deprescribing medications that pose a higher safety burden in older adults (e.g., benzodiazepines). Acute therapies to mitigate anxiety, including benzodiazepines and first-generation antihistamines, are listed on the AGS Beers Criteria® as potentially inappropriate medications in older adults [25]. These treatments should be used sparingly in older adults, generally reserved as short-term bridge treatments to address severe anxiety rather than used as a long-term maintenance treatment. The patient is currently receiving clonazepam and hydroxyzine, two medications indicated to be deprescribed; however, the patient does not have any maintenance treatments to address anxiety symptoms, which would need to be initiated first to avoid the precipitation of worsening anxiety during deprescribing. Given the chronicity of both anxiety and sleep complaints, non-pharmacological strategies are indicated to address both conditions.

Cognitive behavioral therapy, focused on both anxiety and insomnia, could lead to improvements while simultaneously foregoing any safety concerns generated with pharmacotherapy exposure [36, 37]. Although non-pharmacological therapy should be considered first, given the limited risk, the patient may prefer pharmacological approaches solely or in combination with non-pharmacological approaches, which is reasonable to consider. FDA-approved treatments for generalized anxiety disorder (GAD) outside of benzodiazepines include the SSRIs (e.g. escitalopram, paroxetine), the SNRIs (e.g., duloxetine, venlafaxine) and the 5-HT_{1A} partial agonist, buspirone. Although not FDA approved for this indication, sertraline is recognized by treatment guidelines as an effective first-line treatment for GAD [38, 39]. Paroxetine should be avoided in older adults secondary to increased anticholinergic burden [25]. Sertraline, escitalopram, venlafaxine extended-release, and duloxetine are all reasonable first-line options for GAD in older adults [29]. If this patient's blood pressure remains uncontrolled, SNRIs are currently less preferred as hypertension should be controlled prior to initiation. If hypertension is addressed, SNRIs could have a potential advantage in helping reduce pain associated with osteoarthritis [40]. Buspirone is considered a second-line treatment for GAD given its short half-life that requires multiple daily dosing and a poorer response seen in those previously exposed to benzodiazepines [41]. Buspirone can be considered in cases unresponsive to standard treatments or if tolerability concerns arise with SSRI or SNRI use. Buspirone shares many adverse effects like SSRIs including initial activating properties, gastrointestinal disturbances, and headache; however, it does not impact sexual function to the same degree as SSRIs and SNRIs [42, 43].

Pharmacogenomic Considerations

The patient's pharmacogenomic profile has few limitations on initial treatment selection for anxiety; however, it can assist with tailoring the dose titration of the SSRI or SNRI. An SSRI would be a reasonable treatment option to start for this patient, with escitalopram and sertraline being preferred given their indications and low propensity to interact with other his other medications. The patient has a CYP2C19 normal metabolizer phenotype for escitalopram. Given the patient's age, multimorbidity, and polypharmacy, there is no clear reason to consider the CPIC guideline unsafe; therefore, escitalopram can be initiated at its standard starting dose. They also have a CYP2C19 normal and CYP2B6 intermediate metabolizer phenotypes for sertraline. CPIC guidelines (Table 8.1c) and standard approaches in geriatric medicine suggest initiating sertraline at usual starting dose but to titrate slowly to the lowest maintenance dose based on treatment response [23]. If venlafaxine was selected, even though the patient has a CYP2D6 intermediate metabolizer phenotype, no dose adjustments are necessary per the CPIC guidelines (Table 8.1d). However, a 25–50% dose reduction is required for venlafaxine because of the patient's moderate renal dysfunction. Duloxetine is metabolized by CYP1A2 and CYP2D6. Although no pharmacogenomic guidance is available for duloxetine, it would be reasonable to consider its use at a lower dose to mitigate treatment-emergent gastrointestinal side effects [23]. In general, anxiety disorders require lower starting doses of antidepressants to prevent generation of activating side effects, which can worsen anxiety symptoms. Regardless of pharmacogenomic status, it is advised that the lowest starting dose be initiated to prevent treatment-emergent anxiety [44].

Once a therapeutic concentration is achieved for the antidepressant and anxiety is reduced, tapering of the benzodiazepine and hydroxyzine can be started. There is no available pharmacogenomic guidance on tapering clonazepam or hydroxyzine;

however, it is likely best to choose one to taper first. Hydroxyzine has limited potential for dependence or withdrawal, so it can generally be discontinued quickly without a prolonged taper. While benzodiazepine tapering remains a priority for this patient, discontinuing hydroxyzine first may offer a quick initial step within a broader deprescribing strategy. However, the order of discontinuation should be guided by the patient's willingness and preferences. Benzodiazepine tapers should be tailored to the individual needs of the patient as there is not one recommended approach. A general suggestion is to taper by no more than 25% of total daily dose per week for patients who have been receiving long-term treatment with benzodiazepines for greater than one month. As patients reach 50% of their original established dose, the taper may need to be slowed down to reduce precipitated withdrawal reactions [45]. A shared decision-making process, addressing psychological concerns, ameliorating fears of the deprescribing process, and willingness to be flexible during the tapering process are often necessary attributes to improve successful discontinuation.

Overall Recommendations

- The patient is a candidate to start pharmacotherapy for GAD. An SSRI with evidence for treating GAD would be best suited at this time given the patient's mildly elevated blood pressure. Starting either sertraline or escitalopram at their starting doses are reasonable treatments to pursue.
- If sertraline is selected, the maintenance dose will likely need to be lower given the patient's intermediate CYP2B6 metabolizer phenotype.
- A deprescribing plan focusing on tapering the hydroxyzine and benzodiazepine should be introduced and established upfront, with a goal to start the deprescribing process once the antidepressant is optimized. Buy-in and preferences from the patient will help guide selection of which agent to deprescribe first and assist with the tapering speed. The overall tapering plan will not be influenced by the patient's pharmacogenomic panel.
- Regardless of the patient's pharmacogenomic profile, older patients with anxiety disorders should be started on lower antidepressant doses to prevent precipitating activating effects that can worsen anxiety, and inappropriate medications in older adults should be identified for deprescribing.

What Did Pharmacogenomics Change About the Case Outcome?

- Pharmacogenomic testing helped refine maximum dosing thresholds to consider for selecting antidepressants to treat anxiety disorders based on this patient's drug metabolism phenotype.

Case 3

A 68-year-old female presents to her primary care physician's office with worsening symptoms of depression that are now complicated by ruminative thoughts, excessive guilt, and passive suicidal thoughts. She reports being "keyed up" and is now feeling more anxious and worried about her mental wellbeing. She feels that her antidepressant medication is ineffective, and she has had unrelenting gastrointestinal complaints since the dose was increased two weeks ago. Her physician orders a pharmacogenomic test with the results available today for review. The patient is extremely discouraged at both the lack of symptom relief and adverse effects experienced. She is open to exploring a different antidepressant but does not want to take anything that will result in weight gain.

Medical History:

Major depressive disorder
Insomnia
Hypertension
Hyperlipidemia
Obesity categorized by BMI $\geq$ 30

Allergies:

None known

Pertinent Vitals and Labs:

180 lb., 5 ft. 3 in, BMI 31.9, BP 120/80, HR 65

Pharmacogenomic Results:

*CYP2C19 *1/*2* intermediate metabolizer
*CYP2D6 *4/*4* poor metabolizer
*CYP2B6 *6/*6* poor metabolizer

Current Medications:

Atorvastatin 40 mg po daily
Ramipril 10 mg po daily
Amlodipine 10 mg po daily
Ramelteon 8 mg daily at bedtime
Sertraline 100 mg po daily (increased from 50 mg two weeks ago).

Past medications tried and patient reported adverse drug reactions:

Escitalopram—nausea, diarrhea
Fluoxetine—terrible insomnia
Venlafaxine—ineffective, terrible nausea, diarrhea
Citalopram—nausea, anxiety

Clinical (Nongenetic) Considerations

The treatment goals for this patient are to relieve treatment-emergent adverse effects secondary to antidepressant therapy and concomitantly treat the patient's resistant depressive episode. The patient has had several emergent adverse effects secondary to SSRI and SNRI treatment leading to discontinuation. It is unclear the length and duration of the medication trials, and many treatments may have been abandoned prematurely secondary to common adverse effects encountered early in treatment (i.e., gastrointestinal adverse effects). Additional clinical information would help guide whether treatments within the same pharmacological class could be reconsidered, such as another SSRI, SNRI or a retrial of a previous therapy.

SSRIs not previously trialed include paroxetine and fluvoxamine, both of which are less preferred for this patient due to paroxetine's anticholinergic burden and weight-inducing properties, as well as fluvoxamine's potential to interact with CYP1A2 substrates (i.e., ramelteon) and the lack of formal studies on its use for depression in older adults. Duloxetine has evidence for treating depressive episodes in older adults and could be considered as an alternative treatment [29, 46]. Further clarification on the venlafaxine trial could assist in determining desvenlafaxine's potential role as an alternative. If a full trial of venlafaxine was given, there is limited rationale to consider desvenlafaxine as it is venlafaxine's active metabolite and similar effects would be anticipated; however, if venlafaxine was primarily discontinued for adverse effects, desvenlafaxine could serve as an alternative treatment. Bupropion would be a less preferred treatment alternative given the patient's complaints of insomnia and anxiety. While mirtazapine could assist with insomnia symptoms and is safe to use in older adults, its appetite-stimulating effects could be problematic given the patient's desire not to be on therapies that cause weight gain. Vortioxetine could be considered for the patient as it has been formally studied in older adults, with benefits demonstrated in improving processing speed [47]. Other treatments, such as levomilnacipran or vilazodone could be considered, although there are limited trials that formally assess the efficacy of these agents in older adults. Although listed on the AGS Beers Criteria®, TCAs can be considered for refractory forms of depression in older adults given their proven efficacy. TCAs are less tolerable compared to SSRIs; thus, tolerability and premature discontinuation are concerns to consider prior to prescribing.

Therapies typically reserved for refractory depression, defined as the failure of two or more monotherapy treatments with different pharmacologic mechanisms, may also be considered for this patient. However, these treatment strategies are less preferred, as previous pharmacotherapy trials were largely unsuccessful due to adverse effects rather than lack of efficacy. MAOIs are reserved for depressive episodes because of their increased side effect burden and complex drug–drug and drug–food interactions. Similar to concerns with TCAs, MAOIs are not tolerated as well as SSRIs; thus, the likelihood of achieving better tolerability outcomes with these medications is limited [28]. Second-generation antipsychotics have evidence for treating major depressive episodes as augmenting agents, and aripiprazole has established efficacy for depressive episodes in older adults [48]. Second-generation antipsychotics have increased side effect risks in older adults, including inducing extrapyramidal side effects and metabolic complications [49]. Additionally, the patient has not experienced treatment reductions with the current antidepressant, thus switching treatments would be more preferential as antipsychotics have limited efficacy as a monotherapy in treating depressive episodes in older adults [28]. Esketamine is indicated for treatment-resistant depression and could be considered for individuals with failures to standard treatments; however, esketamine failed to achieve statistical significance on its primary outcome in a clinical trial in older adults [50].

Non-pharmacological options should be considered for this patient. Psychotherapy is a recommended intervention for all levels of depression [28, 30]. Electroconvulsive therapy (ECT) is an effective treatment for resistant forms of depression and can be considered for those who have failed multiple treatment trials. Short-term memory impairment is a limitation to treatment and undergoing the procedure can be a barrier for many as it requires undergoing general anesthesia, having adequate transportation to and from the procedure, and living within an area that offers the procedure [51].

Pharmacogenomic Considerations

The patient's pharmacogenomic profile indicates she metabolizes many medications slowly including several antidepressants secondary to having a CYP2B6 poor metabolizer phenotype (impacting sertraline), CYP2C19 intermediate metabolizer phenotype (impacting citalopram, escitalopram, sertraline), and CYP2D6 poor metabolizer phenotype (impacting venlafaxine). In general, reduced metabolism (IM/PM phenotypes) of medications tends to lead to higher plasma concentrations and therefore a higher risk of adverse effects, while increased metabolism (RM/UM phenotypes) tends to lead to lower plasma concentrations and therefore a lower risk of adverse effects, but also lower likelihood of achieving therapeutic efficacy. In this case, the discontinuation of medications was largely due to adverse effects, which tracks quite well with the CYP2B6 PM, CYP2C19 IM, and CYP2D6 PM phenotypes. This patient likely could have benefited from earlier pharmacogenomic testing to help guide dose optimization or avoid therapies that would be difficult to manage given the reduced clearance of several antidepressant medications. Some medications the patient received in the past could be dose adjusted and titrated more prudently based on the patient's drug metabolism status, including sertraline [23]. Given the tolerability concerns experienced with these treatments, the patient may be unwilling to consider using smaller doses of the treatments out of fear of repeatedly experiencing adverse effects. Treatment options that could be considered given the patient's metabolism status include desvenlafaxine, the active metabolite of venlafaxine, which is primarily metabolized via conjugation and should not readily accumulate in the plasma [52]. Both vortioxetine and vilazodone could be considered, but the maximum dose of vortioxetine is 10 mg/day in CYP2D6 poor metabolizer phenotypes [53]. Mirtazapine is metabolized by multiple pathways including CYP3A4, 2D6, and 1A2 and would be minimally impacted by the patient's pharmacogenomic profile, although its adverse effect profile could worsen metabolic comorbidities [54]. Although antipsychotics are a less preferred treatment strategy for this patient, if selected, the patient's pharmacogenomic profile would influence the dosing of two FDA-approved adjunctive antipsychotics for depression: aripiprazole and brexpiprazole. Dosing recommendations informed by pharmacogenomic data for antipsychotics are presented in Tables 8.3a, b and 8.4a, b.

Table 8.3 Dutch Pharmacogenetics Working Group (DPWG) guideline for the gene-drug interaction between CYP2D6, CYP3A4, and CYP1A2 and antipsychotics [56]

(a) DPWG recommendations for aripiprazole, brexpiprazole, haloperidol, risperidone, zuclopenthixol, and CYP2D6

Drug	Phenotype	Recommendation
Aripiprazole	CYP2D6 poor metabolizer	Use no more than 10 mg/day of the oral formulation
Brexpiprazole	CYP2D6 poor metabolizer	Use 50% of the normal dose
Haloperidol	CYP2D6 ultrarapid metabolizer	Use 1.5 times the normal dose or choose an alternative drug not predominantly metabolized by CYP2D6
	CYP2D6 poor metabolizer	Use 60% of the normal dose
Risperidone	CYP2D6 ultrarapid metabolizer	Use an alternative drug
	CYP2D6 poor metabolizer	Use 50–67% of the normal dose
Zuclopenthixol	CYP2D6 intermediate metabolizer	Use 75% of the normal dose
	CYP2D6 poor metabolizer	Use 50% of the normal dose

(b) DPWG recommendation for quetiapine and CYP3A4

CYP3A4 phenotype	Recommendation
Poor metabolizer	For depression: use an alternative drug. For other indications: Use 30% of the normal dose.

Table 8.4 Package labeling recommendations for antipsychotics based on CYP2D6 phenotype

(a) FDA recommendations for pimozide and CYP2D6 [57][a]

CYP2D6 phenotype	FDA label recommendation
Poor metabolizer	For doses above 4 mg daily, CYP2D6 genotyping should be performed. Do not exceed 4 mg/day in CYP2D6 poor metabolizers, and doses should not be increased earlier than 14 days.

(b) FDA recommendations for thioridazine and CYP2D6 [58]

CYP2D6 phenotype	FDA label recommendation
Poor metabolizer	Contraindicated as poor metabolizers have higher systemic concentrations with an increased risk of adverse reactions such as QT prolongation. Data is based on experience with CYP2D6 inhibitors.

[a]DPWG guideline has higher dosing cutoff than FDA label. To follow FDA label for use in the U.S.

Overall Recommendations

- Several options could be considered and discussed with the patient as initial steps, including dose-reducing sertraline (e.g., to 50 mg/day) with a longer treatment duration, or switching medications given the current side effects experienced.
- Patient buy-in will be needed; therefore, there may be skepticism to retry medications previously associated with side effects. Conducting a thorough review of prior antidepressant trials and doses used will be needed before outlining all options and then using shared decision-making, a discussion can be initiated about whether to reintroduce medications like escitalopram or consider alternatives with fewer pharmacogenomic concerns (e.g., vortioxetine or desvenlafaxine).
- Simultaneously, maximizing the role of psychotherapy should also be a priority. Given the anxious features present, lower starting doses and slower titration schedules may help reduce the risk of early discontinuation.
- If symptoms persist after two adequate 12-week treatment trials guided by pharmacogenomic dosing recommendations, or if symptoms worsen significantly, options such as esketamine or ECT are appropriate to consider.

What Did Pharmacogenomics Change About the Case Outcome?

- Although this patient appears to have failed multiple antidepressants, their pharmacogenomic profile suggests that these treatment failures were likely related to tolerability issues secondary from reduced metabolism of several medications, including her current regimen.
- Pharmacogenomic testing provided a rationale for tolerability concerns with multiple antidepressants, stemming from slower metabolism and decreased clearance of common first-line antidepressants.
- Pharmacogenomic testing helped identify therapeutic alternatives to consider for this patient with the best safety profile considering her biologic age, multimorbidity, and polypharmacy. If it was done earlier, it could potentially have prevented the ruling out of several previous therapies as dosing strategies could be applied based on patient's drug metabolism phenotype. The patient's frustration and anxiety may also have been mitigated.
- Irrespective of the patient's pharmacogenomic profile, non-pharmacologic options such as psychotherapy should always be considered for patients with depression if readily available.

Conclusion

Pharmacogenomic testing serves as an additional decision-support tool to help guide clinicians managing and prescribing pharmacotherapy for depression and anxiety in older adults. Although pharmacogenomic testing is not a universal remedy for solving medication failures associated with depression and other psychiatric disorders, it could serve as a useful tool to help medication selection and optimize dosing of many psychotropic medications, rule out potentially problematic therapies, and help improve the safety and tolerability of treatments prescribed in a population that is more vulnerable to CNS-related adverse effects.

References

1. Reynolds K, Pietrzak RH, El-Gabalawy R, Mackenzie CS, Sareen J. Prevalence of psychiatric disorders in U.S. older adults: findings from a nationally representative survey. World Psychiatry. 2015;14(1):74–81.
2. Zenebe Y, Akele B, Mulugeta WS, Necho M. Prevalence and determinants of depression among old age: a systematic review and meta-analysis. Ann Gen Psychiatry. 2021;20(1):55.
3. Thakur M, Blazer DG. Depression in long-term care. J Am Med Dir Assoc. 2008;9(2):82–7.
4. Kok RM, Reynolds CF 3rd. Management of Depression in older adults: a review. JAMA. 2017;317(20):2114–22.

5. Sekhon S, Patel J, Sapra A. Late-Life Depression [Updated 2023 Jun 26]. In: StatPearls [Internet]. Treasure Island (FL): StatPearls Publishing; 2025 Jan-. Available from: https://www.ncbi.nlm.nih.gov/books/NBK551507/.
6. Beekman AT, Deeg DJ, Braam AW, Smit JH, Van Tilburg W. Consequences of major and minor depression in later life: a study of disability, well-being and service utilization. Psychol Med. 1997;27(6):1397–409.
7. Cuijpers P, Vogelzangs N, Twisk J, Kleiboer A, Li J, Penninx BW. Comprehensive meta-analysis of excess mortality in depression in the general community versus patients with specific illnesses. Am J Psychiatry. 2014;171(4):453–62.
8. Wei J, Hou R, Zhang X, Xu H, Xie L, Chandrasekar EK, et al. The association of late-life depression with all-cause and cardiovascular mortality among community-dwelling older adults: systematic review and meta-analysis. Br J Psychiatry. 2019;215(2):449–55.
9. Shahal T, Davidow J, Gatchel J, Forester B. Geriatric mood disorders. In: Sadock B, Sadock VA, Ruiz P, editors. Kaplan and Sadock's comprehensive textbook of psychiatry. 11th ed. Philadelphia: Wolters Kluwer Health; 2024.
10. Depression in adults: treatment and management. London: National Institute for Health and Care Excellence (NICE); 2022 Jun 29. (NICE Guideline, No. 222.). Available from: https://www.ncbi.nlm.nih.gov/books/NBK583074/.
11. Fässberg MM, Cheung G, Canetto SS, Erlangsen A, Lapierre S, Lindner R, et al. A systematic review of physical illness, functional disability, and suicidal behaviour among older adults. Aging Ment Health. 2016;20(2):166–94.
12. Gournellis R, Lykouras L, Fortos A, Oulis P, Roumbos V, Christodoulou GN. Psychotic (delusional) major depression in late life: a clinical study. Int J Geriatr Psychiatry. 2001;16(11):1085–91.
13. Szymkowicz SM, Gerlach AR, Homiack D, Taylor WD. Biological factors influencing depression in later life: role of aging processes and treatment implications. Transl Psychiatry. 2023;13(1):160.
14. Cole MG, Dendukuri N. Risk factors for depression among elderly community subjects: a systematic review and meta-analysis. Am J Psychiatry. 2003;160(6):1147–56.
15. Fiske A, Wetherell JL, Gatz M. Depression in older adults. Annu Rev Clin Psychol. 2009;5:363–89.
16. Vink D, Aartsen MJ, Schoevers RA. Risk factors for anxiety and depression in the elderly: a review. J Affect Disord. 2008;106(1–2):29–44.
17. Casey DA. Depression in older adults: a treatable medical condition. Prim Care. 2017;44(3):499–510.
18. Husain-Krautter S, Ellison JM. Late life depression: the essentials and the essential distinctions. Focus (Am Psychiatr Publ). 2021;19(3):282–93.
19. Wojt IR, Cairns R, Clough AJ, Tan ECK. The prevalence and characteristics of psychotropic-related hospitalizations in older people: a systematic review and meta-analysis. J Am Med Dir Assoc. 2021;22(6):1206–14.e5.
20. Bousman CA, Bengesser SA, Aitchison KJ, Amare AT, Aschauer H, Baune BT, et al. Review and consensus on Pharmacogenomic testing in psychiatry. Pharmacopsychiatry. 2021;54(1):5–17.
21. Bousman CA, Oomen A, Jessel CD, Tampi RR, Forester BP, Eyre HA, et al. Perspectives on the clinical use of pharmacogenetic testing in late-life mental healthcare: a survey of the American Association of Geriatric Psychiatry Membership. Am J Geriatr Psychiatry. 2022;30(5):560–71.
22. Marshe VS, Islam F, Maciukiewicz M, Bousman C, Eyre HA, Lavretsky H, et al. Pharmacogenetic implications for antidepressant pharmacotherapy in late-life depression: a systematic review of the literature for response, pharmacokinetics and adverse drug reactions. Am J Geriatr Psychiatry. 2020;28(6):609–29.
23. Bousman CA, Stevenson JM, Ramsey LB, Sangkuhl K, Hicks JK, Strawn JR, et al. Clinical Pharmacogenetics implementation consortium (CPIC) guideline for CYP2D6, CYP2C19,

CYP2B6, SLC6A4, and HTR2A genotypes and serotonin reuptake inhibitor antidepressants. Clin Pharmacol Ther. 2023;114(1):51–68.
24. International Society of Psychiatric Genetics. Genetic Testing Statement. [Internet]. International Society of Psychiatric Genetics; [updated 2019 Mar 11; cited 2025 Aug 20].. Available from: https://ispg.net/genetic-testing-statement/.
25. American Geriatrics Society 2023 updated AGS Beers Criteria® for potentially inappropriate medication use in older adults. J Am Geriatr Soc. 2023;71(7):2052–81.
26. Cicali EJ, Elchynski AL, Cook KJ, Houder JT, Thomas CD, Smith DM, et al. How to integrate CYP2D6 Phenoconversion into clinical Pharmacogenetics: a tutorial. Clin Pharmacol Ther. 2021;110(3):677–87.
27. Cicali EJ, Wiisanen K. The importance of phenoconversion when using the CYP2D6 genotype in clinical practice. Pharmacogenomics. 2022;23(14):749–52.
28. Lam RW, Kennedy SH, Adams C, Bahji A, Beaulieu S, Bhat V, et al. Canadian Network for Mood and Anxiety Treatments (CANMAT) 2023 Update on Clinical Guidelines for Management of Major Depressive Disorder in Adults: Réseau canadien pour les traitements de l'humeur et de l'anxiété (CANMAT) 2023 : Mise à jour des lignes directrices cliniques pour la prise en charge du trouble dépressif majeur chez les adultes. Can J Psychiatr. 2024;69(9):641–87.
29. MacQueen GM, Frey BN, Ismail Z, Jaworska N, Steiner M, Lieshout RJ, et al. Canadian Network for Mood and Anxiety Treatments (CANMAT) 2016 clinical guidelines for the management of adults with major depressive disorder: section 6. Special populations: youth, women, and the elderly. Can J Psychiatry. 2016;61(9):588–603.
30. Malhi GS, Bell E, Bassett D, Boyce P, Bryant R, Hazell P, et al. The 2020 Royal Australian and New Zealand College of Psychiatrists clinical practice guidelines for mood disorders. Aust N Z J Psychiatry. 2021;55(1):7–117.
31. Weihs KL, Settle EC Jr, Batey SR, Houser TL, Donahue RM, Ascher JA. Bupropion sustained release versus paroxetine for the treatment of depression in the elderly. J Clin Psychiatry. 2000;61(3):196–202.
32. Schatzberg AF, Kremer C, Rodrigues HE, Murphy GM Jr. Double-blind, randomized comparison of mirtazapine and paroxetine in elderly depressed patients. Am J Geriatr Psychiatry. 2002;10(5):541–50.
33. Bondareff W, Alpert M, Friedhoff AJ, Richter EM, Clary CM, Batzar E. Comparison of sertraline and nortriptyline in the treatment of major depressive disorder in late life. Am J Psychiatry. 2000;157(5):729–36.
34. Beach SR, Celano CM, Noseworthy PA, Januzzi JL, Huffman JC. QTc prolongation, torsades de pointes, and psychotropic medications. Psychosomatics. 2013;54(1):1–13.
35. Allergan USA, Inc Celexa (citalopram hydrobromide) [package insert]. Madison: Allergan USA, Inc; 2022.
36. Hall J, Kellett S, Berrios R, Bains MK, Scott S. Efficacy of cognitive behavioral therapy for generalized anxiety disorder in older adults: systematic review, meta-analysis, and meta-regression. Am J Geriatr Psychiatry. 2016;24(11):1063–73.
37. Hertenstein E, Trinca E, Wunderlin M, Schneider CL, Züst MA, Fehér KD, et al. Cognitive behavioral therapy for insomnia in patients with mental disorders and comorbid insomnia: a systematic review and meta-analysis. Sleep Med Rev. 2022;62:101597.
38. Andrews G, Bell C, Boyce P, Gale C, Lampe L, Marwat O, et al. Royal Australian and New Zealand College of Psychiatrists clinical practice guidelines for the treatment of panic disorder, social anxiety disorder and generalised anxiety disorder. Australian New Zealand J Psychiatry. 2018;52(12):1109–72.
39. Bandelow B, Allgulander C, Baldwin DS, Costa D, Denys D, Dilbaz N, et al. World Federation of Societies of Biological Psychiatry (WFSBP) guidelines for treatment of anxiety, obsessive-compulsive and posttraumatic stress disorders – version 3. Part I: anxiety disorders. World J Biol Psychiatry. 2023;24(2):79–117.
40. Osani MC, Bannuru RR. Efficacy and safety of duloxetine in osteoarthritis: a systematic review and meta-analysis. Korean J Intern Med. 2019;34(5):966–73.

41. Schatzberg AF, DeBattista C. Anxiety Disorders In: Schatzberg's manual of clinical psychopharmacology. 9th ed. Washington, DC: American Psychiatric Association Publishing; 2019.
42. Othmer E, Othmer SC. Effect of buspirone on sexual dysfunction in patients with generalized anxiety disorder. J Clin Psychiatry. 1987;48(5):201–3.
43. Landén M, Eriksson E, Agren H, Fahlén T. Effect of buspirone on sexual dysfunction in depressed patients treated with selective serotonin reuptake inhibitors. J Clin Psychopharmacol. 1999;19(3):268–71.
44. Bandelow B, Michaelis S, Wedekind D. Treatment of anxiety disorders. Dialogues Clin Neurosci. 2017;19(2):93–107.
45. Ogbonna CI, Lembke A. Tapering patients off of benzodiazepines. Am Fam Physician. 2017;96(9):606–10.
46. Raskin J, Wiltse CG, Siegal A, Sheikh J, Xu J, Dinkel JJ, et al. Efficacy of duloxetine on cognition, depression, and pain in elderly patients with major depressive disorder: an 8-week, double-blind, placebo-controlled trial. Am J Psychiatry. 2007;164(6):900–9.
47. Katona C, Hansen T, Olsen CK. A randomized, double-blind, placebo-controlled, duloxetine-referenced, fixed-dose study comparing the efficacy and safety of Lu AA21004 in elderly patients with major depressive disorder. Int Clin Psychopharmacol. 2012;27(4):215–23.
48. Lenze EJ, Mulsant BH, Roose SP, Lavretsky H, Reynolds CF 3rd, Blumberger DM, et al. Antidepressant augmentation versus switch in treatment-resistant geriatric depression. N Engl J Med. 2023;388(12):1067–79.
49. Behrman S, Burgess J, Topiwala A. Prescribing antipsychotics in older people: a mini-review. Maturitas. 2018;116:8–10.
50. Ochs-Ross R, Daly EJ, Zhang Y, Lane R, Lim P, Morrison RL, et al. Efficacy and safety of Esketamine nasal spray plus an oral antidepressant in elderly patients with treatment-resistant depression-TRANSFORM-3. Am J Geriatr Psychiatry. 2020;28(2):121–41.
51. Wilkinson ST, Kitay BM, Harper A, Rhee TG, Sint K, Ghosh A, et al. Barriers to the implementation of Electroconvulsive Therapy (ECT): results from a Nationwide survey of ECT practitioners. Psychiatr Serv. 2021;72(7):752–7.
52. Sopko MA Jr, Ehret MJ, Grgas M. Desvenlafaxine: another "me too" drug? Ann Pharmacother. 2008;42(10):1439–46.
53. Trintellix (vortioxetine) [package insert]. Lexington: Takeda Pharmaceuticals America, Inc.; 2023.
54. Anttila SA, Leinonen EV. A review of the pharmacological and clinical profile of mirtazapine. CNS Drug Rev. 2001;7(3):249–64.
55. Hicks JK, Sangkuhl K, Swen JJ, Ellingrod VL, Müller DJ, Shimoda K, et al. Clinical pharmacogenetics implementation consortium guideline (CPIC) for CYP2D6 and CYP2C19 genotypes and dosing of tricyclic antidepressants: 2016 update. Clin Pharmacol Ther. 2017;102(1):37–44.
56. Beunk L, Nijenhuis M, Soree B, de Boer-Veger NJ, Buunk AM, Guchelaar HJ, et al. Dutch Pharmacogenetics Working Group (DPWG) guideline for the gene-drug interaction between CYP2D6, CYP3A4 and CYP1A2 and antipsychotics. Eur J Hum Genet. 2024;32(3):278–85.
57. Pimozide. Package insert. Endo USA, INC; 2017. Accessed June 14, 2025.
58. Thioridazine. Package insert. Chartwell RX, LLC; 2025. Accessed June 14, 2025.

Chapter 9
Bringing It All Together

Yee Ming Lee, Teresa DeLellis, Stephanie Sibicky, and David Bright

Introduction

This final chapter offers an opportunity to synthesize and apply the geriatric pharmacogenomics (PGx) principles thus far outlined in somewhat organ-system-specific contexts. The exercise aims to reinforce the information in Section I and provides a final chance to practice the application skills from Section II in a holistic patient scenario.

Case

A 77-year-old female is seen for a six-month outpatient follow-up appointment. While she has a long history of hypertension and diabetes, she has been relatively well controlled until this visit. Her concerns include constantly being tired and "sick of everything hurting all of the time." She says that her frustration with "feeling

Y. M. Lee (✉)
Associate Professor of Clinical Pharmacy, Skaggs School of Pharmacy and Pharmaceutical Sciences, University of Colorado, Aurora, CO, USA
e-mail: yeeming.lee@cuanschutz.edu

T. DeLellis
School of Pharmacy, College of Health Professions, Nursing and Pharmacy, Manchester University, Fort Wayne, IN, USA

S. Sibicky
Clinical Professor and Assistant Dean for Academic Affairs and Assessment, Bouvé College of Health Sciences, Northeastern University, Boston, MA, USA

D. Bright
Professor of Pharmaceutical Sciences, College of Pharmacy, Ferris State University, Big Rapids, MI, USA

T. DeLellis et al. (eds.), *Applying Pharmacogenomics to Optimize Geriatric Care*, https://doi.org/10.1007/978-3-032-10653-7_9

achy all over," osteoarthritis (knee), and her neuropathy has caused her to lose motivation to be as active as she was even just six months ago. She has stopped walking her dog twice daily and has gained 15 pounds since her last visit. When asked about this shift, her responses triggered the physician to perform an assessment of depression, and an associated desire to start an antidepressant at this time.

You recommend obtaining a PGx test and explain its purpose. The patient recalls getting PGx testing a decade ago that showed she has a CYP2C19 rapid metabolizer phenotype. You explain that a panel test with more genes, is more cost-effective than getting one enzyme at a time. The patient agreed to the test, and her results are provided below.

Medical History

Hypertension (37 years)
Dyslipidemia (6 months ago)
Type 2 diabetes (18 years) with neuropathy
Osteoarthritis (8 years)

Allergies:

No known allergies

Pertinent Vitals and Labs:

191 lb., 5 ft. 3 in, BP 144/84, HR 78, eGFR 44 ml/min/1.73m^2
Lipid panel (six months ago): TG 230 mg/dL, HDL 44 mg/dL, LDL 122 mg/dL
Hgb A1c 7.6% today (7.1% 6 months ago)

PGx Results:

*CYP2C19*1/*17* rapid metabolizer
*CYP2D6 *4/*5* poor metabolizer
*CYP2C9 *2/*3* poor metabolizer
*SLCO1B1 *5/*5* poor function
ABCG2 Normal function

Current Medications:

Aspirin 81 mg by mouth daily
Atorvastatin 40 mg by mouth daily
Metoprolol succinate 100 mg by mouth daily
Celecoxib 200 mg by mouth daily
Metformin 1000 mg by mouth twice daily

Clinical (Nongenetic) Considerations

The primary goal for this patient is pain control to restore function and quality of life. Her pain is multifactorial, and the resulting loss of functionality is likely at least partially responsible for exacerbation of other chronic conditions and a new diagnosis of depression.

A thorough evaluation of pain symptoms must be conducted to assist in identifying the most likely precipitators of the recent increase in pain. In any new onset of symptoms in older adults, medications should be evaluated as causative agents first to avoid a prescribing cascade. Although she has been stable on atorvastatin for a few months, statin-induced myalgia needs to be evaluated and ruled out as a cause, especially considering the consistency and widespread nature of the pain. This patient presents with multicomplexity and recent advancement in conditions, which could precipitate physiological changes and interactions leading to atorvastatin accumulation and resulting adverse effects even without dosing changes. Something as simple as introducing consistent grapefruit juice into her diet could have caused this. If the statin is felt to be a likely cause, shared decision-making should take place regarding the need for continued statin use. Evidence of benefit is less strong when used for primary prevention and in persons over 75 years of age. While the last LDL was not extremely elevated, diabetes does increase cardiovascular risk. It would be reasonable to obtain a CAC score and consider the lack of life-limiting illness in this patient. Aspirin use has similar recommendations and should be included in the conversation (see Chap. 3).

Osteoarthritis is another likely cause of pain. The frequency of celecoxib use should be ascertained, since this medication can increase cardiovascular risk, contribute to fluid retention and exacerbation of hypertension, and increase the risk of kidney injury in combination with other risk factors present such as hypertension and diabetes. Topical therapies are first-line, and topical NSAIDs such as diclofenac have demonstrated similar efficacy in knee osteoarthritis compared to systemic NSAIDs without carrying the same risk for gastrointestinal bleeding or other adverse effects (see Chap. 1). Another option could be to use a scheduled dose of 1000 mg of acetaminophen three times a day, barring no liver disease or excessive alcohol use, which could help with underlying pain control.

Neuropathy, another possible source of increased pain, is not currently being treated. Of the possible treatments for neuropathic pain, discussed in more detail in Chap. 1, duloxetine is an attractive option since it could also help treat this patient's newly diagnosed depression. There is also some evidence of efficacy for duloxetine in the treatment of osteoarthritis, particularly with comorbid mental illness.

Other opportunities for therapy optimization unrelated to pain include the treatment of diabetes and hypertension. Use of metformin or GLP-1 receptor antagonists is recommended regardless of hemoglobin A1c in patients who are overweight and/or have high ASCVD risk, and either would be reasonable to initiate in this patient. The hemoglobin A1c increase is likely related to recent inactivity related to pain and depressive symptoms and may be reversed without further need for medications if those conditions are adequately controlled. Her blood pressure is above goal at this visit, which could be exacerbated due to pain. However, metoprolol is no longer a first-line antihypertensive, and decreased efficacy has been demonstrated with aging. Other reasons this is not an ideal choice for hypertension treatment include the potential to mask symptoms of hypoglycemia and an increased risk of orthostatic hypotension and depression. Preferred agents in patients with diabetes include ACE inhibitors or ARBs since they have renal protective effects and will control blood pressure better than a selective beta-blocker. Changing from metoprolol monotherapy to an ACEi or ARB monotherapy, while treating pain, could be enough to achieve her blood pressure goal.

Pharmacogenomic Considerations

The insidious onset of new adverse effects presenting from a previously stabilized medication is much more likely related to medications for which an individual is already genetically predisposed to adverse effects. This patient's phenotype of *SLCO1B1* poor function increases the suspicion that atorvastatin could be contributing to the widespread constant pain. CPIC guidelines indicate that a high SAMs risk is associated with atorvastatin doses from 40 to 80 mg (see Chap. 3). If statin therapy is still desired after the shared decision-making discussion, the patient's *ABCG2* normal function would indicate rosuvastatin is the safer alternative. However, if the benefit of continuing the statin therapy does not outweigh the risk, PGx information could strengthen the rationale to discontinue statin therapy for this patient.

The patient's other CYP phenotypes also point to duloxetine as a good choice to treat both pain and depression. Many other antidepressants are likely to increase the risk of accumulation and adverse effects (*CYP2D6* poor metabolizer phenotype) or require high maintenance doses to produce the desired effect (*CYP2C19* rapid metabolizer phenotype). Current CPIC guidelines have no recommendations for duloxetine dosing, as current data show no clinically meaningful impact of *CYP2D6* on duloxetine (see Chap. 4).

The PGx panel shows an incidental finding of *CYP2C9* poor metabolizer phenotype, which furthers the case against systemic use of *CYP2C9*-impacted NSAIDs. This affects both her current medications of celecoxib and ibuprofen, a potential over-the-counter self-treatment; as both will have reduced clearance. CPIC guidelines recommend using 50% of the lowest starting dose and titrating to 50% of the maximum dose (see Chap. 1). If the patient were to self-treat pain with over-the-counter analgesics as her pain regimen changes, she should be counseled to utilize naproxen rather than ibuprofen, since naproxen is not impacted by *CYP2C9*.

The patient's metoprolol, although already recommended as a change, is metabolized by CYP2D6, and PGx guidance does exist for this medication. CPIC guidelines suggest that after five days of use, if the patient's blood pressure, heart rate, and other clinical features indicate that the drug is working as appropriate without any significant safety concerns or adverse effects, no further dose or medication adjustment is necessary (see Chap. 3).

Summary of Recommendations

- Treat pain and depression first with topical diclofenac, acetaminophen, and duloxetine. If breakthrough systemic NSAID therapy is needed, naproxen would be safer than ibuprofen or celecoxib based on PGx.
- Adequate treatment of pain and depression should bring back activity and help to lower hemoglobin A1c within goal. Regardless, other medications could be considered with shared decision-making, for cardiovascular protection.
- It is probably atorvastatin therapy that is causing pain in addition to osteoarthritis and neuropathy.
- To review statin and aspirin therapy through shared decision-making discussion. If statin is continued, rosuvastatin would be a reasonable alternative statin.

What Did PGx Change About This Case Outcome?

- Knowledge of the patient's *SLCO1B1* phenotype helped solidify the reality that her previously stable atorvastatin dose could be causing new adverse effects and provided guidance on an alternative less likely to cause the same problems.
- The patient's *CYP2C9* phenotype supported the discontinuation of celecoxib and the avoidance of ibuprofen, facilitating what is often a challenging treatment plan for patients to accept based on clinical nongenetic data alone. It also helped guide to the safest option, if the patient persisted in desiring a back-up systemic NSAID.
- With many antidepressant therapy options to choose from and some conflicting evidence of benefit for co-morbidities, the patient's PGx narrowed down duloxetine as the best option.
- While PGx guidance exists for some diabetes and antihypertensive medications, it did not change the treatment recommendation.

Index

T. DeLellis et al. (eds.), *Applying Pharmacogenomics to Optimize Geriatric Care*, https://doi.org/10.1007/978-3-032-10653-7

MIX
Papier aus verantwortungsvollen Quellen
Paper from responsible sources
FSC® C105338

If you have any concerns about our products,
you can contact us on
ProductSafety@springernature.com

In case Publisher is established outside the EU,
the EU authorized representative is:
Springer Nature Customer Service Center GmbH
Europaplatz 3, 69115 Heidelberg, Germany

Printed by Libri Plureos GmbH
in Hamburg, Germany